Who's Who
in Fluorescence 2003

Who's Who in Fluorescence 2003

Who's Who
in Fluorescence 2003

Edited by

Chris D. Geddes and Joseph R. Lakowicz
Center for Fluorescence Spectroscopy
Baltimore, Maryland

Springer Science+Business Media, LLC

Library of Congress Cataloging-in-Publication Data

ISBN 978-0-306-47689-1   ISBN 978-1-4615-0173-2 (eBook)
DOI 10.1007/978-1-4615-0173-2

© 2003 Springer Science+Business Media New York
Originally published by Kluwer Academic/Plenum Publishers, New York in 2003

To the Fluorescence Community

with thanks

Chris D. Geddes and Joseph R. Lakowicz

Preface

Chris D. Geddes, Ph.D.

When we made the transition of the editor-in-chief position of the *Journal of Fluorescence* from Joseph Lakowicz to Chris D. Geddes, in November 2001, we both already had many ideas of how the *Journal* could further be moulded to fulfil the continuously changing needs of the Fluorescence community. Fluorescence had changed from the *Principles of Fluorescence* driven research, to an interdisciplinary science that was application and biotechnology driven. The expansion of fluorescence not only meant that the number of workers employing fluorescence had substantially risen in the past 10 years, but workers are now more widespread throughout the world, instead of in the traditional fluorescence hotspots, such as in the labs which pioneered developments in fluorescence spectroscopy. It was therefore challenging to invite a new *Journal of Fluorescence* editorial board, which reflects/ed these changes, and the current state-of-play in fluorescence.

Joseph R. Lakowicz, Ph.D.

However, we soon realised that the diversity of fluorescence was such that it could not be fully reflected by 30 or so *Journal of Fluorescence* editorial board members, but in fact, could only be reflected by fluorescence workers, whether, academic, industry or government, themselves. Subsequently the idea of the *Who's Who in Fluorescence* Annual volume was born, and, we are now pleased to present the first volume. Many e-mail invitations were sent to fluorescence workers and companies around the World. The response to the Who's *Who in Fluorescence 2003* invitations were excellent. The volume not only reflects the breadth of fluorescence, by the speciality descriptions provided by contributors, but it also provides complete contact details, key publications as well as contributor speciality keywords. We hope that you will find the volume both useful and informative, and we look forward to receiving your contributions and suggestions for the *Who's Who in Fluorescence 2004* volume. Further details and instructions for authors can be found at http://cfs.umbi.umd.edu/

In closing we would also like to thank the fluorescence instrumentation companies for their support, many of which have entries towards the back of the volume, as well as the fluorescence workers themselves, some 312 personal entries appear in the 2003 volume.

Finally a special thanks goes to Caroleann Aitken and Mary Rosenfeld for their help in compiling the volume.

We thank you all for supporting the *Who's Who in Fluorescence*.

Chris D. Geddes and Joseph R. Lakowicz

About the Editors.

Joseph R. Lakowicz Ph.D., is director of the Center for Fluorescence Spectroscopy at the University of Maryland School of Medicine in Baltimore. He has a BS in chemistry from LaSalle University and an MS and Ph.D. in Biochemistry from the University of Illinois at Urbana. He is also the founding editor of the *Journal of Biomedical Optics* and the *Journal of Fluorescence* and Co-President of the *Society of Fluorescence*.

Chris D. Geddes Ph.D., is an assistant professor at the, Center for Fluorescence Spectroscopy and Director of the Institute of Fluorescence, at the University of Maryland Biotechnology Institute, in Baltimore. He has a B.Sc. from Lancaster University in England and a Ph.D. in physical chemistry (fluorescence spectroscopy) from the University of Wales Swansea. He is the editor of the *Journal of Fluorescence*, an editorial board member of *Dyes and Pigments* and Executive Director of the new *Society of Fluorescence*.

Contents

Personal Entries

Company Entries

Date submitted: 13th September 2002 **Omoefe O. Abugo, Ph.D.**

Unison Technology Services,
WorldCom Inc., 1201 Seven Locks Rd,
Rockville, MD 20854,
U.S.A.
Tel: 301 6106184
oabugo@maranatha.net

Specialty Keywords: Lifetime Fluorescence, Polarization, Blood, Tissues, Drug Compliance.

Research has involved the use of fluorescence techniques in the detection and characterization of fluorescing moieties in biological systems and for biophysical measurements. Examples include the use of modulation and polarization sensing methods in the detection of red and NIR emitting dyes in tissue and whole blood. Other examples include the use of frequency domain fluorometry and FRET for the study of interactions between hemoglobin and proteins on the cytoplasmic domain of red cell membranes.

O.O. Abugo, Z. Gryczynski and J.R. Lakowicz (1999). J. Biomed. Optics, 4, 429-442.
O.O. Abugo, R. Nair and J.R. Lakowicz (2000). Anal. Biochem.., 279, 142-150.

Date submitted: 9th September 2002 **Ramesh C. Ahuja, Ph.D.**

LaVision GMBH,
Anna-Vandenhoeck-Ring 19,
D37081 Goettingen,
Germany.
Tel: +49 551 900 40 / Fax: +49 551 900 4100
Info@lavision.de
www.lavision.de

WE COUNT ON PHOTONS

Specialty Keywords: Time-Resolved Imaging, Spectroscopy and Microscopy, FLIM, FRET, Microscopy.

Current Status: Manager of optical diagnostics (based on reflection, fluorescence, multiphoton, SHG, time-gated Raman, imaging spectroscopy) products for Materials and Life Sciences.
LaVision specializes in offering *state-of-the-art* intensified, gated (down to 50 ps) or gain-modulated (up to 1GHz) CCD-based cameras as well as customer-specific turn-key imaging spectroscopy and microscopy systems incorporating e.g. camera, laser, aux. light source, coupling optics, imaging spectrograph, sample chamber or microscope, peripheral devices and data analysis software. Integrated turnkey systems are offered for: fluorescence microscopy, time-resolved spectroscopy, fluorescence lifetime imaging microscopy, multifocal multiphoton microscopy, combustion, flow (Micro-PIV) and spray (drop size, distribution) diagnostics.

Date submitted: 10th August 2002

Engin U. Akkaya, Ph.D.

Department of Chemistry,
Middle East Technical University,
06531 Ankara,
Turkey.
Tel: 90 312 210 5126   Fax: 90 312 210 1280
akkayaeu@metu.edu.tr

Specialty Keywords: Fluorescent chemosensors, Molecular logic gates, Molecular devices.

Current interests: Design and synthesis of novel fluorogenic and chromogenic chemosensors for cations, anions and carbohydrates. Novel sensing schemes. Calixarene-based ion-pair sensors and allosterical modulation of binding interactions. Oxidative PET and cation/anion modulation of oxidative PET. Antenna systems. Diazapyrenium-based fluorescent pseudorotaxanes. Novel and efficient sensitizers for photodynamic therapy.   Fluorescent chemosensors for dopamine.

C. N. Baki and Engin U. Akkaya (2001). Boradiazaindacene appended calix[4]arene: Fluorescence sensing of pH near neutrality, *J. Org. Chem*. **66**, 1512-1513.
B. Turfan and Engin U. Akkaya (2002). Modulation of Boradiazaindacene Emission by Cation Mediated Oxidative PET, *Organic Lett*. **4**, 2857-2859.

Date submitted: 16th May 2002

Fabrizio Alfieri, (Ph.D. Student)

Institute of Protein Biochemistry,
Via Pietro Castellino, 111
Naples, 80131,
Italy.
Tel: +39 0816132312   Fax: +39 0816132270

alfier@dafne.ibpe.na.cnr.it

Specialty Keywords: Biosensor, Fluorescence, Thermophilic enzymes.

My scientific interests deal with the development of innovative protein biosensor for analytes of high clinical, environment and food interests based on the utilization of enzymes and proteins isolated from mesophilic and thermophilic organisms.
My primary goal is to contribute to the realization of new methods for analytes sensing using fluorescence techniques.
In this regard, my Ph.D. thesis is focused on the development of a thermostable and non-consuming substrate fluorescence biosensor for glucose.

Date submitted: 28th August 2002

Annette Alfsen, Ph.D., M.D.

Institut Cochin, IFR 116, Inserm, CNRS,
Université René Descartes,
22, rue Méchain, 75014 - Paris, France.
Tel: (331) 40 51 64 45   Fax: (331) 40 51 64 54
alfsen@cochin.inserm.fr

Specialty Keywords: Cell biology - Biophysics-Cell membranes.

My field of research has been centered on the molecular interactions at the biophysical level and in cell integrated structures. The physical chemistry of cell membranes and interaction with the surrounding medium and with the neighboring cells is still under study for the entry of HIV in epithelial cells and the infection processes.

Alfsen A. 1989 Membrane dynamics and Molecular traffic and sorting in mammalian cells. Prog. Biophys. Mol. Biol. 54: 145-57
Alfsen A. and Bomsel M. 2002. HIV-1 gp 41 Envelope Residues 650-685 Exposed on Native Virus act as a Lectin to bind Epithelial Cell GalactosylCeramide. J. Biol. Chem. 277 .25649-659.

Date submitted: 12th September 2002

Wajih Al-Soufi, Ph.D.

Universidad de Santiago de Compostela,
Facultad de Ciencias, Departamento de Química Física,
Campus Universitario s/n, E-27002 Lugo,
Spain.
Tel: +34 982 223325   Fax: +34 982 22 49 04
alsoufi@lugo.usc.es

Specialty Keywords: Fluorescence, Data analysis.

Current interests: Study of the influence of confined media such as cyclodextrins on proton transfer and charge transfer processes. Design of fluorescent probes for the characterisation of supramolecular structures formed by cyclodextrins. Development and implementation of new data analysis methods for steady state and time resolved fluorescence data.

W. Al-Soufi, M. Novo y M. Mosquera (2001). Principal Component Global Analysis of fluorescence and absorption spectra of 2-(2'-hydroxyphenyl)benzimidazole. *Appl. Spectrosc.,* **55**, 630-636. E. Alvarez-Parrilla, W. Al-Soufi, P. Ramos Cabrer, M. Novo y J. Vázquez Tato(2001). Resolution of the association equilibria of 2-(p-toluidinyl)-naphthalene-6-sulfonate (TNS) with cyclodextrin and a charged derivative. *J. Phys. Chem. B,* **105**, 5994-6003.

Date submitted: 31st August 2002

Limburgs
Universitair Centrum

transnationale Universiteit
Limburg
Belgium/the Netherlands

Marcel M. Ameloot, Ph.D.

Dept. MBW,
Limburgs Universitair Centrum,
Universitaire Campus,
B-3590 Diepenbeek, Belgium.
Tel: + 32 (0)11 268 502   Fax: + 32 (0) 11 268 599
Marcel.Ameloot@luc.ac.be
www.luc.ac.be/biomed

Specialty Keywords: Microfluorimetry, Time-resolved fluorescence, Data analysis.

The research deals with the application of steady-state and time-resolved (imaging) microfluorimetry in cell physiology and in the development of biosensors. Currently the focus is on the behavior of oligodendrocytes and the myelin membrane within the framework of *multiple sclerosis* research.

S. Despa, J. Vecer, P. Steels and M. Ameloot (2000) Lifetime-based fluorescence microscopy of the ion indicator Sodium Green in HeLa cells *Anal. Biochem.* **281**, 159-175.
N. Boens, J.P. Szubiakowski, E. Novikov and M. Ameloot (2000) Testing the identifiability of a model for reversible intermolecular two-state excited-state processes *J. Chem. Phys.* **112**, 8260-8266

Date submitted: 8th August 2002

John E. Anderson, Ph.D.

U.S. Army Engineering Research and Development Center,
Fluorescence Remote Sensing Lab,
USAERDC-TEC, 7701 Telegraph Road, Alexandria,
Virginia, 22315, USA.
Tel: 703 428 6698   Fax: 703 428 8176
John.Anderson@erdc.usace.army.mil
www.tec.army.mil

Specialty Keywords: Fluorescence Remote Sensing, Enzyme Substrates, Waterborne pathogens.

Dr. Anderson's research interests involve active and passive fluorescence sensing to detect and identify waterborne pathogens. Both biotic (defined substrates) and abiotic (polymers) strategies are used with novel bioreporters to recover signatures relevant to pathogenic activity. A major research goal is the molecular-level characterization of relevant fluorophores scaled to the imaging domain for synoptic representation.

Anderson, J.E., Webb, S.R., Fischer, R.L., Smith, C.B., Dennis, J.R., and Di Benedetto, J. (2002). *In situ* detection of the pathogen indicator *E. coli* using active laser-induced fluorescence imaging and defined substrate conversion. *Journal of Fluorescence* (12) 1 p. 51-55.

Date submitted: 27th August 2002

David L. Andrews, Ph.D.

School of Chemical Sciences,
University of East Anglia,
Norwich, NR4 7TJ,
U.K.
Tel: +44 1603 592014   Fax: +44 1603 59203
david.andrews@physics.org
www.uea.ac.uk/~c051

Specialty Keywords: Quantum Electrodynamics, Resonance Energy Transfer, Nonlinear Optics.

Andrews's research centers on molecular photophysics. Early work led to the unified theory of energy transfer,[1] subsequently eliciting electronic effects of the condensed phase through a polariton formulation. His group was first to identify and predict the characteristics of two-photon resonance energy transfer, anticipating experiments on biological systems. Recent work begun in 1998[2] focuses on energy pooling in optically nonlinear photoactive systems.

D.L. Andrews (1989). A unified theory of radiative and radiationless molecular energy transfer *Chem. Phys.* **135**, 195-201.
R.D. Jenkins and D.L. Andrews (1998). Three-center systems for energy pooling: quantum electrodynamical theory *J. Phys. Chem. A.* **102**, 10834-10842.

Date submitted: 15th August 2002

Pavel Anzenbacher, Ph.D., D.Sc.

Inst. of Pharmacology, Faculty of Medicine,
Palacky University,
Hnevotinska st. 3, Olomouc,
CZ-779 00, Czech Republic.
Tel: +420 58 563 2569   Fax: +420 58 563 2966
anzen@tunw.upol.cz

Specialty Keywords: Protein conformation, Tryptophans, Heme enzymes.

Active sites of cytochromes P450 and other heme enzymes differ in amino acid residues to reflect their function and specificity. Tryptophan fluorescence is studied by stationary approach as well as by time-resolved techniques. Interaction with enzyme substrates often produce fluorescence changes which are characteristic for different cytochrome P450 enzymes. FCS gives then information on changes in protein aggregation and overall conformation.
R. Lange, Anzenbacher P., Müller S., Maurin L., Balny C. (1994) Interact. of tryptophan residues in cytochrome P450scc with a fluorescence quencher *Eur.J.Biochem.* 226, 963-970
Bemeš M., Hudeček J., Anzenbacher P., Anzenbacher P., Hof M. (2001) Coumarin 6, resorufins and flavins: Suitable chromophores for FCS of biol. molecules.*Coll.Czech.Chem.Commun.*66, 855-869

Arden-Jacob, J.
Bagatolli, L.A.

Date submitted: 20th September 2002

Date submitted: 20th September 2002

Date submitted: 20th September 2002

Jutta Arden-Jacob, Ph.D.

ATTO-TEC GmbH,
57008 Siegen,
Germany.
Tel: +49 (0) 271 7404735
Arden-jacob@atto-tec.de
www.atto-tec.com

Specialty Keywords: Fluorescent dyes, Biolabelling, Red-absorbing chromophors.

My research is focused on the chemical synthesis and characterization of new red-absorbing fluorophors. I am particularly interested in new fluorescent dyes which are suitable for biolabelling.

Arden-Jacob J., Frantzeskos J., Kemnitzer N. U., Zilles A., Drexhage K.H., Spectrochim. Acta 57A, 2271-2283 (2001).
Arden-Jacob J., Frantzeskos J., Kemnitzer N. U., Zilles A., Drexhage K.H., J. Fluoresc. 7, 91S-93S (1997) .

Date submitted: 7th August 2002

Luis A. Bagatolli, Ph.D.

Memphys - Center for Biomembrane Physics,
Department of Physics, University of Southern Denmark,
Campusvej 55, DK-5230 Odense M,
Denmark.
Tel: +45 65 50 34 76 Fax: +45 66 15 87 60
bagatolli@memphys.sdu.dk
www.memphys.sdu.dk/

Specialty Keywords: Multiphoton microscopy, Polarity sensitive probes, Lipid / lipid and Lipid / protein interactions.

My primary research goal is to study lipid/lipid and lipid protein interactions in natural and model membranes. The fluorescence parameters measured in traditional experiments involving liposome solutions can be measured at the level of single vesicles using fluorescence microscopy. Using this last approach is possible to establish a correlation between the microscopic organization on the surface of single vesicles with the physical parameters determined at molecular level on the lipid bilayer (lipid mobility, lipid hydration, etc).

Bagatolli L.A. and E. Gratton. (2001) *J. of Fluorescence* 11:141-160.
Sanchez S., L. A. Bagatolli, E. Gratton, T. Hazlett (2002) *Biophys. J.* 82:2232-2243.

Date submitted: 26th August 2002

Gary A. Baker, Ph.D.

Bioscience Division, Los Alamos National Laboratory,
Mailstop J586,
Los Alamos, NM 87545,
USA.
Tel: 505 665 6910
gabaker@lanl.gov

Specialty Keywords: Bio–fluorescence, Organized assemblies, Non-aqueous enzymology.

Current research interests include: two-photon imaging, heme proteins, spectroelectrochemistry, mesoporous materials, DNA folding, ionic liquids.

G.A. Baker et al., "Extending the Reach of Clinical Assays to Optically-Dense Specimens by Using Two-Photon Excited Fluorescence Polarization," *Anal. Chem.* **2000**, *72*, 5748.
G.A. Baker et al., "Effects of Fluorescent Probe Structure on the Dynamics at Cysteine-34 within Bovine Serum Albumin: Evidence for Probe-Dependent Modulation of the Cybotactic Region," *Biopolymers* **2001**, *59*, 502.

Date submitted: 26th August 2002

Sheila N. Baker, Ph.D.

Chemistry Division, Los Alamos National Laboratory,
Mailstop J514,
Los Alamos, NM 87545,
USA.
Tel: 505 667 2658
sbaker@lanl.gov

Specialty Keywords: Ionic liquids, Supercritical fluids, Electroluminescence.

Current research interests include: ionic liquids, 'green' chemistry, emulsions in CO_2, beryllium toxicity and bio–interactions, novel fluorine chemistry.

S.N. Baker et al., "The Cybotactic Region Surrounding Fluorescent Probes Dissolved in 1-Butyl-3-Methylimidazolium Hexafluorophosphate: Effects of Temperature and Added Carbon Dioxide," *J. Phys. Chem. B* **2001**, *105*, 9663.
S.N. Baker et al., "Temperature-Dependent Microscopic Solvent Properties of 1-Butyl-3-Methylimidazolium Hexafluorophosphate: Correlation with $E_T(30)$ and Kamlet-Taft Polarity Scales," *Green Chem.* **2002**, *4*, 165.

Date submitted: 22nd August 2002

Aleksander Balter, Ph.D.

Institute of Physics, N. Copernicus University,
Grudziadzka 5,
87-100 Torun,
Poland.
Tel: +48 (56) 6113216   Fax: +48 (56) 6225397
balter@phys.uni.torun.pl

Specialty Keywords: Molecular biophysics, Photoluminescence, Sonoluminescence.

Current interests: Photophysical and photochemical properties of fluorescence probes. Fluorescence and Raman spectroscopy of protein-water interactions. Single bubble sonoluminescence.

A. Kamińska, M. Kowalska and A. Balter (1999). A comparative study of the effect of exogenous and endogenous photostabilizers in the lens crystallin photodegradation, *J.Fluorescence* **9,** 213-219.
J. Szubiakowski, A. Balter, W. Nowak, K. Wisniewski and K. Aleksandrzak (1999) Substituent-sensitive anisotropic rotations of 9-acetoxy-10-phenylanthracenes. Fluorescence anisotropy decay and quantum-mechanical study, *Chem. Phys. Lett.* **313**, 473-483.

Date submitted: 4th September 2002

Susan L. Bane, Ph.D.

Department of Chemistry,
State University of New York at Binghamton,
Binghamton,
New York 13902, USA.
Tel: (607) 777 2927   Fax: (607) 777 4478
sbane@binghamton.edu
chemistry.binghamton.edu/BANE/bane.html

Specialty Keywords: Microtubules, Ligand/receptor interactions, New fluorescent probes.

We are interested in determining the molecular mechanisms by which antimicrotubule drugs (such as paclitaxel (Taxol), colchicine, vinblastine, and combretastatin) interact with the protein tubulin and with microtubules. We use a variety of fluorescence spectroscopy techniques to elucidate these mechanisms. Design and synthesis of new fluorescent probes is also in progress.
Baloglu, E., Kingston, D. G. I., Patel, P., Chatterjee, S. K. and Bane, S. L. (2001) Synthesis and microtubule binding of fluorescent paclitaxel derivatives. *Bioorg. Med. Chem. Lett.* **11**, 2249-2252.
Han, Y., Malak, H., Chaudhary, A. G., Chordia, M. D., Kingston, D. G. I., and Bane, S. (1998) Distances between the paclitaxel, colchicine and exchangeable GTP binding sites on tubulin. *Biochemistry* **37**, 6636-6644.

Date submitted: 13th September 2002

Elisabeth Bardez, Ph.D.

Conservatoire National des Arts et Métiers,
292, rue Saint-Martin,
5141 Paris Cedex 03,
France.
Tel: +33 (0)1 40 27 25 92   Fax: +33 (0)1 40 27 23 62
bardez@cnam.fr

Specialty Keywords: Excited-state proton transfer, Fluorescent sensors for aluminum(III), Photoinduced tautomerization.

Current interests: Photoinduced tautomerization in amphoterous bifunctional compounds (hydroxyquinolines, hydroxycoumarins). Photoinduced proton ejection from dihydroxynaphthalenes. Design of hexadentate fluorogenic ligands for aluminum determination includind bidentate sub-units as 8-hydroxyquinoline, chromotropic acid, etc.

E. Bardez et al. (2001). From 8-hydroxy-5-sulfoquinoline to new related fluorogenic ligands for complexation of aluminium(III) and gallium(III). *New J. Chem.* **25**, 1269 - 1280.

E. Bardez (1999). Excited-state proton transfer in bifunctional compounds *Israel J. Chem.* **39**, 319 - 332.

Date submitted: 10th July 2002

George Barisas, Ph.D.

Department of Chemistry,
Colorado State University,
Fort Collins, CO 80523,
USA.
Tel: 970 491 6641   Fax: 970 491 1801
barisas@lamar.colostate.edu

Specialty Keywords: Cell, Membrane, Dyna.

We examine the dynamics and distributions of cell surface molecules in relation to membrane signal transduction events in cells of the immune system and in gonadotropin-responsive cells. We measure lateral motions through photobleaching recovery and single-particle tracking, rotational motions through time-resolved phosphorescence anisotropy and fluorescence depletion anisotropy and spacial distributions through fluorescence resonant energy transfer and photoproximity labeling. We have developed new or improved implementations of each of the above techniques.

Date submitted: 9th September 2002

Grzegorz Bartosz, Ph.D.

Department of Molecular Biophysics, University of Łódź,
Banacha 12/16, Łódź,
PL 90-237,
Poland.
Tel: +48 42 6354476   Fax: +48 42 6354473
gbartosz@biol.uni.lodz.pl
www.biol.uni.lodz.pl/~kbm

Specialty Keywords: Reactive oxygen species, Transport, Membrane fluidit.

Membrane fluidity estimated with fluorescent probes and spin labels; fluorimetric and spin trap detection of reactive oxygen species; fluorimetric assays of total antioxidant capacity and cell survival; flow cytometric studies of apoptosis, fluorimetric studies of transport (mainly by Multidrug Resistance Proteins).

Grzelak A, Rychlik B, Bartosz G.:Light-dependent generation of reactive oxygen species in cell culture media. Free Radic Biol Med. 30:1418-425 (2001).
Jakubowski W, Bartosz G.:Estimation of oxidative stress in Saccharomyces cerevisae with fluorescent probes. Int J Biochem Cell Biol. 29:1297-1230 (1997).

Date submitted: 21st August 2002

Joseph M. Beechem, Ph.D.

Molecular Probes, Inc.,
Dept. of BioSciences,
4849 Pitchford Ave,
Eugene, OR 97402 9165.
Tel: (541) 242 0435   Fax: (541) 984 5698
joe@probes.com

Specialty Keywords: Assays, Kinetics, Proteomics, Genomics, imaging.

My research focuses on the development of fluorescence-based technologies/tools in order to solve biomedically relevant problems. Research emphasis integrates the (supposedly) disparate technologies of: proteomics, genomics, high-throughput screening, microarrays, and high-resolution *ex-vivo* and *in-vivo* imaging. Emphasis is placed on obtaining multiplexed correlated kinetic data using multiple detection devices (e.g., microscopes, microplate readers, mass-specs, 2-D gels, microarrays, etc.) during physiological transitions. Currently, fluorescence technology is the only approach that has the inherent dynamic-range, sensitivity, and timing-resolution to span such a wide range of applications.

W. F. Patton and J. M. Beechem. "Rainbow's end: the quest for multiplexed fluorescence quantitative analysis in proteomics." *Curr. Opin. Chem. Biol.*, 6(1):63-69 (2002).
Beechem, J. M. (1992) Global analysis of Biophysical Data. *Methods in Enzymology 210*, 37-54.

Date submitted: 30ᵗʰ August 2002

Martin J. Behne, M.D.

Department of Dermatology,
University of California San Francisco,
4150 Clement Street, San Francisco,
CA 94121, USA.
Tel: (415) 750 2091   Fax: (415) 751 3927
behnemj@itsa.ucsf.edu

Specialty Keywords: FLIM, Ion gradients, Epidermis.

The physiologic roles and effects of specific ion gradients, and the transporters that generate these gradients are the focus of my interest. With biochemical, molecular, and microscopic methods their expression in epidermal keratinocytes and in whole epidermis is investigated. In whole epidermis, fluorescence lifetime imaging is used to visualize the gradients generated and/or maintained by such transporters, and to further elucidate the spatio-temporal changes in these gradients, their functions, and effects in epidermal differentiation, homeostasis, and disease.

K. M. Hanson, M. J. Behne, N. P. Barry, T. M. Mauro, E. Gratton, and R. M. Clegg (2002)., Two-Photon Fluorescence Lifetime Imaging of the Skin Stratum Corneum pH Gradient *Biophys J* **83**(3), 1682-1690.

Date submitted: 8ᵗʰ August 2002

Kevin D. Belfield, Ph.D.

Department of Chemistry and School of Optics / CREOL,
University of Central Florida,
4000 Central Florida Blvd., Orlando, 32816,
USA.
Tel: 407 823 1028   Fax: 407 823 2252
kbelfiel@mail.ucf.edu
www.cas.ucf.edu/chemistry/personnel/belfield.html

Specialty Keywords: Two-photon photochemistry, Microfabrication, Non-destructive imaging.

Molecular structure/linear absorption/nonlinear absorption relationships of organic molecules, the development of highly efficient two-photon fluorescent dyes, and two-photon polymerization and photochromism are being investigated.

K.D. Belfield, M.V. Bondar, O.V. Przhonska and K.J. Schafer (2002). Steady-State Spectroscopic and Fluorescence Lifetime Measurements of New Two-Photon Absorbing Fluorene Derivatives *J. Fluorescence* **12**, in press.

K.D. Belfield and K.J. Schafer (2002). A New Photosensitive Polymeric Material for Optical Data Storage using Multichannel Two-Photon Fluorescence Readout *Chem. Mater.* **14**, in press.

Berberan-Santos, M.N.
Beuthan, J.

Date submitted: 31st August 2002

Mário N. Berberan-Santos, Ph.D.

Centro de Química-Física Molecular,
Instituto Superior Técnico,
1049-001 Lisboa,
Portugal.
Tel: +351 218419254 Fax: +351 218464455
berberan@ist.utl.pt

Specialty Keywords: Photophysical kinetics, Resonance energy transfer, Multichromophoric systems.

Current interests: Photophysics of fullerenes (early work included discovery of thermally activated delayed fluorescence[1]). Radiative transport in scattering media (previous work on combined radiative and nonradiative transport[2] included development of a stochastic theory and its experimental test). Excitation energy hopping and transfer in multichromophoric systems.

[1]M.N. Berberan-Santos and J.M.M. Garcia (1996). Unusually strong delayed fluorescence of C_{70}, *J. Am. Chem. Soc.* **118**, 9391-9394.
[2]M.N. Berberan-Santos, E.N. Pereira, and J.G. Martinho (1999), Dynamics of radiative transport, in *Resonance Energy Transfer*, D.L. Andrews and A.A. Demidov eds., Wiley, Chichester.

Date submitted: 20th June 2002

Jürgen Beuthan, Ph.D.

Institute for Medical Physics,
Freie Universität Berlin
Fabeckstr. 60-62, 14195 Berlin, Germany.
Tel: +49 30 84492323 Fax: +49 30 84494377
j.beuthan@lmtb.de
www.fu-berlin.de

Specialty Keywords: Optical Biopsy, Cell metabolism, Medical applications.

Current Research Interests: My research is focused on advancing fluorescence applications in medicine using native autofluorescence compounds like NADH and Cytokeratin. These investigations are carried out both time-resolved and in cw mode. They serve for investigating metabolic changes, such as cancer or ischemia, using optical methods.

J Beuthan, O. Minet, G. Müller (1993): Observations of the fluorescence response of the coenzyme NADH in biological samples. *Opt. Lett.,* **18,** 1098-1100.
J. Beuthan, O. Minet, G. Müller (1998): Optical Biopsy of Cytokeratin and NADH in the Tumor Border Zone. *Annals New York Academy Sciences,* **838**, 150-170.

Date submitted: 27th August 2002

Kankan Bhattacharyya, Ph.D.

Department of Physical Chemistry,
Indian Association for the Cultivation of Science,
Jadavpur, Kolkata 700 032,
India.
Tel: (91) 33 473 3542   Fax: (91) 33 473 2805
pckb@mahendra.iacs.res.in
www.iacs.res.in/pckb.html

Specialty Keywords: Ultrafast dynamics, Organized assembly.

Our major interest is to study dynamics in organized assemblies using time resolved fluorescence spectroscopy. Solvation dynamics, proton/electron transfer, isomerization and orientational dynamics are found to be dramatically retarded in micelles & reverse micelles, lipids, cyclodextrin, protein, zeolite etc. For instance, solvation dynamics of water in an organized assembly displays a component 100-1000 times slower than that in bulk water.

K. Bhattacharyya, and B. Bagchi (2000) *J. Phys. Chem. A* **104**, 10603.
K. Bhattacharyya (2001) *J. Fluorescence* **11**, 167.

Date submitted: 6th September 2002

Jan G. Bieschke, Ph.D.

Institute for Neuropathology,
Ludwid-Maximilians-University,
Marchioninistr. 17, D-81377 Munich,
Germany.
Tel: +49 89 7095 7912   Fax: +49 89 7095 4903
bieschke@lmu.de

Specialty Keywords: FCS, Protein misfolding, Single molecules.

We study aggregation processes in neurodegenerative diseases caused by protein misfolding on a single molecule level. Our aim is the characterization of intermediate steps in aggregation and detection and characterization of misfolded protein aggregates in diagnostic applications by multi-color confocal fluorescent spectroscopy. Systems examined include PrP (Prion diseases), Aß (Alzheimer's disease) and synuclein (Parkinson's disease).

Bieschke J, Giese A, Schulz-Schaeffer W, Zerr I, Poser S, Eigen M, and Kretzschmar H (2000) Ultrasensitive detection of pathological prion protein aggregates by dual-color scanning for intensely fluorescent targets. Proc. Natl. Acad. Sci. U. S. A **97**, 5468-5473.

Birmingham, J.J.
Borie, C.

Date submitted: 29th August 2002

John J. Birmingham, Ph.D.

Unilever Research Port Sunlight,
Quarry Road East, Bebington,
Wirral, Merseyside CH63 3JW,
United Kingdom.
Tel: (0151) 641 3351   Fax: (0151) 641 1841
John.Birmingham@unilever.com

Specialty Keywords: Photobleaching, Lifetime imaging.

Research emphasis on development of fluorescence technologies to aid detection and imaging of industrially relevant ingredients deposited on both natural and man-made surfaces at low levels from consumer products. Key techniques include fluorescence photobleaching methods (time and frequency domains) and nanosecond timescale lifetime imaging, the latter implemented in the frequency domain for both widefield imaging and laser scanning geometries to suit a range of distance scales from microscopic to large macroscopic.

J.J.Birmingham (1997) *J.Fluorescence* **7**(1):45-54.
J.J.Birmingham (1999) in A.Kotyk (ed) , *Fluorescence Microscopy and Fluorescent Probes 3*, Espero, Prague, pp.23-35.

Date submitted: 5th August 2002

Christophe Borie.

Aventis Pharama,
Assay Development, HTS,
13 quai J. Guesde
Vitry 94400,
France.
Tel: 33 1 58 93 30 63
Christophe.borie@aventis.com

Specialty Keywords: HTS, Assay development, HTRF.

The use of fluorescence in my activity is directed around two principal axes: on the one hand the use of the transfer of fluorescence in time resolved for the biochemical assays in homogeneous phase, on the other hand cells based assays with use of Acumen technology (scanner laser beam).

Date submitted: 17th July 2002

Guido Böse, Ph.D.

Experimental Biophysics,
MPI Biophysical Chemistry,
Am Faßberg 11, Göttingen, 37077,
Germany.
Tel: +49 551 201 1380   Fax: +49 551 201 1435
gboese@gwdg.de

Specialty Keywords: FCS, DNA repair, RNA interference.

Fluorescence Correlation Spectroscopy is a versatile tool for the examination of biomolecules concerning binding and conformational changes. In my DNA repair project UvrAB are examined for DNA binding and conformational changes with dual color crosscorrelation analysis and single molecule FRET measurements.
In the RNA interference project fluorescently labelled RNAs are used for FCS measurements while silencing gene expression.

Microplate Enzyme-Linked Immunosorbent Assay for the Detection of Primary DNA Alterations Based on the Interaction with UvrA/ UvrB, Böse et al. (2001), Anal. Biochem. 292, 1-7.

Date submitted: 5th September 2002

Ludwig Brand, Ph.D.

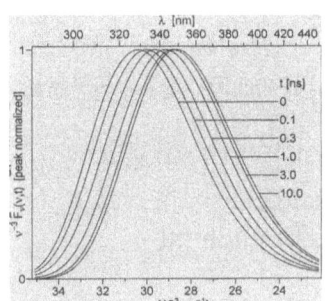

Biology Department, Johns Hopkins University,
3400 North Charles Street,
Baltimore, MD 21218,
USA.
Tel: 410 515 7298   Fax: 410 516 7298
Ludwig.Brand@jhu.edu

Specialty Keywords: Fluorescence, Proteins, Membranes.

The interest of our laboratory is to understand the static and dynamic structure of proteins, biological membranes, and nucleic acids. The work includes studies of the interactions between macromolecules and the relation between structure and function. A variety of excited-state processes such as proton transfer, energy transfer, exciplex and excimer formation and solvent relaxation are being investigated so that these processes can be better used to study biological macromolecules in vivo and in vitro.

Toptygin, D. Savichenko, R.S., Meadow, N.D. and Brand, L.,"Homogeneous Spectrally and Time-Resolved Fluorescence Emission from Single-Tryptophan of IIA^{Glc} Protein.", Journal of Physical Chemistry B, 105, 2043-2055 (2001).

Braut–Boucher, F.
Bright, F.V.

Date submitted: 27th August 2002

Françoise Braut – Boucher, Ph.D.

**Institut national
de la santé et de la recherche
médicale**

Phagocytes et Réponses Inflammatoires, Unité INSERM 479,
Faculté de Médecine Xavier Bichat, 16 rue Henri Huchard,
75018 Paris Cedex,
France.
Tel: 01 42 86 20 80   Fax: 01 42 86 20 80
francoise.braut@biomedicale.univ-paris5.fr

Specialty Keywords: Fluorescence microplate assays,
Oxidative agression, Cellular interactions.

Modifications of cellular adhesive capacities associated with oxidative agression are implicated in several pathologies: cardiovascular diseases, inflammation and metastasis. The consequences of induced-oxidative stress on cellular interactions are studied on different models *in vitro*. Besides immunological methods or flow cytometry, fluorescence microplate assays are performed using specific fluorescent probes. Cell adhesion (1), reactive oxygen species production, intracellular thiols (2) and apoptosis are analysed in relation to the expression of adhesive molecules.

Braut-Boucher F, Pichon J, Rat P, Adolphe M, Aubery M, Font J. J Immunol Methods, 1995, **178**, 41-51.
Plantin-Carrenard E, Braut-Boucher F, Bernard M, Derappe C, Foglietti M.J, Aubery M. Journal of Fluorescence, 2000, **10**, 167-173.

Date submitted: 25th May 2002

Frank V. Bright, Ph.D.

Department of Chemistry, 511 Natural Sciences Complex,
University at Buffalo, The State University of New York
Buffalo, NY 14260-3000, USA.
Tel: 716 645 6800 ext. 2162   Fax: 716 645 6963
chefvb@acsu.buffalo.edu
www.chem.buffalo.edu/Bright.html

Specialty Keywords: Analytical chemistry, Biosensors, New materials, Proteins in restricted space, Supercritical fluids.

Research efforts focus on: biomolecule dynamics at/in surfaces and chemical biosensing, photophysics within microheterogeneous systems, developing and tailoring sol-gel-derived composites as new sensing platforms, new materials for tracheal wound repair, supercritical fluid science and technology, multi-photon excitation strategies in chemical analysis, and laser-based chemical instrumentation.

M.A. Kane, S. Pandey, G.A. Baker, S.A. Perez, E.J. Bukowski, D.C. Hoth and F.V. Bright, "Effects of Density on the Intramolecular Hydrogen Bonding, Tail-Tail Cyclization, and Mean-Free Tail-to-Tail Distances of Pyrene End-Labeled Poly(dimethylsiloxane) Oligomers Dissolved in Supercritical CO_2," *Macromolecules* **2001**, *34*, 6831-6838.

Date submitted: 13th August 2002

Rasmus Bro, Ph.D.

Dept. of Dairy and Food Science, The Royal Vet. & Agri. Uni,
Rolighedsvej 30, Frederiksberg,
Denmark, 1958,
Denmark.
Tel: +45 35283296   Fax: +45 35283245
rb@kvl.dk
www.models.kvl.dk

Specialty Keywords: Chemometrics, Multi-way analysis.

R. Bro works on dedicated chemometric models for fluorescence EEM data especially within food analysis and medicine. New multi-way models can be used to separate mixture measurements into the underlying fluorophores thus enabling quantitative and qualitative analysis. The mathematical chromatography made possible by multi-way analysis is a new and potentially very interesting technique in complicated problems.

R. Bro (1997). PARAFAC. Tutorial and applications Chemom Intell Lab Syst, **38**, 149-171.
R. Bro (1998). Multi-way Analysis in the Food Industry. Ph.D. thesis, University of Amsterdam (NL), www.mli.kvl.dk/staff/foodtech/brothesis.pdf.

Date submitted: 23rd August 2002

Jean-Claude Brochon, Ph.D.

L.B.P.A., Ecole Normale Supérieure de Cachan, C.N.R.S.,
61, avenue du Président Wilson,
94235 Cachan Cedex,
France.
Tel: +33 (0) 1 47 40 27 17   Fax: +33 (0) 1 47 40 24 79
brochon@lbpa.ens-cachan.fr
www.lbpa.ens-cachan.fr/photobm/

Specialty Keywords: Proteins, Time-resolved anisotropy, Data analysis.

Structural dynamics and function of biological macromolecules from time-resolved fluorescence *in vitro*. Currently, protein dynamics, self-assembly of proteins, protein-nucleic acids and protein-protein interactions. A recent project in my laboratory is to extend these studies, *in vivo*, in using 2-photons confocal microscopy and FLIM techniques; application to retrovirus replication. High hydrostatic pressure for study of protein plasticity. Application of the Maximum Entropy Method of data analysis in time-resolved spectroscopies.

Deprez, E., Tauc, P., Leh, H., Mouscadet, J-F., Auclair, C. Hawkins, M. E., Brochon, J-C., DNA binding induces dissociation of the multimeric form of HIV-1 integrase : A time-resolved fluorescence anisotropy study, Proc. Nat. Acad. Sci. USA, (2001) 98, 10090- 10095.

Date submitted: 21st August 2002

Peter J. Butler, Ph.D.

Bioengineering, The Pennsylvania State University,
228 Hallowell Building, University Park,
Centre County, 16802,
USA.
Tel: 814 865 8086   Fax: 814 863 049
pjbbio@engr.psu.edu
www.bioe.psu.edu

Specialty Keywords: Vascular biology, Endothelial cells, Mechanotransduction, Spectroscopy.

Our laboratory is interested in applying sophisticated imaging techniques including confocal microscopy and time correlated, single photon counting spectroscopy to study the effects of mechanical forces (e.g. fluid shear stress) on the mechanics and dynamics of molecules in living cells and tissues involved in mechanotransduction. We wish to use these techniques to understand the molecular bases of mechanically-induced changes in vascular biology.

Date submitted: 29th July 2002

Elisabete M. Castanheira, Ph.D.

Departamento de Física, Universidade do Minho,
Campus de Gualtar,
4710-057 Braga,
Portugal.
Tel: + 351 253 604321 Fax: +351 253 678981
ecoutinho@fisica.uminho.pt
www.fisica.uminho.pt

Specialty Keywords: Molecular spectroscopy, Biophysics, Microheterogeneous systems.

Current interests: Fluorescent probes; self-assembly molecules; biocompatible colloids (structure and applications); polymer photophysics; dynamics of macromolecules; microaggregates (structure and applications); molecular spectroscopy; kinetics.

G. Hungerford, E.M.S. Castanheira, M.E.C.D. Real Oliveira, M.G. Miguel, H.D. Burrows (2002) Monitoring ternary systems of $C_{12}E_5$/water/tetradecane via the fluorescence of solvatochromic probes, *J. Phys. Chem. B* **106**, 4061-4069.
M.E.C.D. Real Oliveira, G. Hungerford, E.M.S. Castanheira, M.G. Miguel, H.D. Burrows (2000) Monitoring the phase transition of $C_{12}E_5$/water/alkane microemulsions through excimer formation, *J. Fluorescence* **10**, 347-353.

Date submitted: 31st August 2002

Miguel A. R. B. Castanho, Ph.D.

Dept of Chemistry and Biochemistry,
University of Lisbon, Campo Grande Ed. C8,
Lisbon, PT 1749-016,
Portugal.
Tel: + 351 21 7500931   Fax: + 351 21 7500088
Castanho@fc.ul.pt
www.dqb.fc.ul.pt/docentes/mcastanho

Specialty Keywords: Biomembrane, Quenching, Structure.

Fluorescence spectroscopy is used to obtain structural information on the organization of polyene antibiotics and peptides in aqueous media and lipidic bilayers. The agreement between experimental data and theoretical expectations in different techniques (e.g., quenching, energy transfer and migration, anisotropy and linear dichroism), leads to conclusions about, for instance, partition coefficients, aggregation, location, orientation and lateral and rotational dynamics of probes. Recently, the experimental results have been compared to predictions obtained by brownian dynamics simulations.

Date submitted: 11th September 2002

Zoran G. Cerovic, D.Sc.

LURE-CNRS,
Bât 203, Orsay,
F-91898,
France.
Tel: +33 (0) 164468209   Fax: +33 (0) 164464148
zoran.cerovic@lure.u-psud.fr
www.lure.u-psud.fr

Specialty Keywords: Photosynthesis, Chlorophyll, Polyphenols.
Studies on the interactions between photochemistry and biochemistry in photosynthesis. Spectroscopy of functional intact isolated chloroplasts and reconstituted chloroplast systems. Investigations on the origin of variable chlorophyll fluorescence *in vivo*. Time–resolved measurements of fluorescence (sub–nanosecond). Investigation on the origin of blue-green fluorescence of plants, and on the UV-excited fluorescence of leaves in general. Design of fluorescence signatures for remote sensing of vegetation.
Latouche, G., Cerovic, Z.G., Montagnini, F. & Moya, I. (2000) Light-induced changes of NADPH fluorescence in isolated chloroplasts: a spectral and fluorescence lifetime study. *Biochim. Biophys. Acta*, **1460**(2-3): 311-329. Ref 2: Ounis, A., Cerovic, Z.G., Briantais, J.-M. & Moya, I. (2001) Dual excitation FLIDAR for the estimation of epidermal UV absorption in leaves and canopies. *Remote Sens. Environ.*, **76**: 33-48.

Date submitted: 27th August 2002

Abhijit Chakrabarti, Ph.D.

Biophysics Division, Saha Institute of Nuclear Physics,
37 Belgachia Road,
Kolkata, 700037,
India.
Fax: +91 33 337 4637
abhijit@biop.saha.ernet.in
www.saha.ernet.in

Specialty Keywords: Spectrin, Membrane, Local anesthetics.

My major interest is in the study of the membrane skeletal spectrin self-assembly of erythrocytes probing the Trp's in tailor made membranes and micelles using steady state and time-resolved fluorescence. I have also been working on the polarity estimate and localization of the unique Prodan/Pyrene-binding sites in spectrin. My group is also engaged in the study of membrane localization of the local anesthetics dibucaine and tetracaine in presence of cholesterol and the glycoshingolipid, GM1.

M Mandal, K Mukhopadhyay, S Basak and A Chakrabarti (2001). Biochim. Biophys. Acta 1511, 146-155.
S Ray and A Chakrabarti (2002). Cell Motil. Cytoskeleton. *in press*.

Date submitted: 27th July 2002

Philip J. Chan, Ph.D., HCLD

Gynecology / Obstetrics, Loma Linda University,
11370 Anderson St., Loma Linda,
California, 92354,
U.S.A.
Tel: 909 558 2851   Fax: 909 558 2450
pchann@yahoo.com
www.llu.edu/lluhc/fertility

Specialty Keywords: Andrology, Embryos, Infertility.

Formerly at the Comparative Medicine Study Section NIH and presently an inspector for the College of American Pathologists, my research is in fluorescent assay development, sorting for gender selection, HPV transgenesis, gene mutations, apoptosis in gametes and embryos.

Chan PJ, Mann SL, Corselli JU, Patton WC, King A, Jacobson JD. A simple DNA disc chip in a microarray design based on comparative genomic hybridization for sperm DNA analysis. Fertil Steril 2002;77:1056-1059.
Lee CA, Huang CTF, King A, Chan PJ. Differential effects of human papillomavirus DNA types on p53 tumor-suppressor gene apoptosis in sperm. Gynecol Oncol 2002;85:511-516.

Date submitted: 13th September 2002

Lin L. Chandler, Ph.D.

SPEX Fluorescence, Jobin Yvon, Inc.,
3880 Park Ave, Edison,
NJ, 08820-3012,
USA.
Tel: 732 494 8660 ext 236
Lin_Chandler@jyhoriba.com

Specialty Keywords: Anisotropy, Photon-counting, Frequency-domain.

Member of a team of scientists providing fluorescence applications support, training and new methods development for users of SPEX spectrofluorometers. Support is provided for all users interested in applying high sensitivity photon-counting, steady-state fluorescence spectroscopy, fluorescence microscopy and picosecond time-resolved, frequency-domain methods to their own research projects.

Date submitted: 18th June 2002

Amitabha Chattopadhyay, Ph.D.

Centre for Cellular & Molecular Biology,
Uppal Road, Hyderabad 11794,
India.
Tel: +91 40 719 2578   Fax: +91 40 716 0311
amit@ccmb.res.in
www.ccmb..res.in

Specialty Keywords: Fluorescence spectroscopy and quenching, Solvent relaxation, Biomembranes and micelles.

My major research interest is the application of fluorescence spectroscopic approaches to problems in membrane and receptor biology. We have successfully used approaches based on slow solvent relaxation rates in organized molecular assemblies such as membranes and micelles to address key issues about their organization. Another area of interest is the application of fluorescence techniques to understand membrane receptor organization and dynamics.
A. Chattopadhyay, and S. Mukherjee (1999) Depth-dependent solvent relaxation in membranes: wavelength-selective fluorescence as a membrane dipstick *Langmuir* **15**, 2142-2148.
A. Chattopadhyay, and S. Mukherjee (1999) Red edge excitation shift of a deeply embedded membrane probe: implications in water penetration in the bilayer *J. Phys. Chem. B* **103**, 8180-8185.

Date submitted: 1st August 2002

Alex F. Chen, M.D., Ph.D.

Department of Pharmacology and Toxicology & the
Neuroscience Program,
Michigan State University,
B403 Life Sciences Building,
East Lansing, MI 48824-1317, USA.
Tel: 517 432 2730   Fax: 517 353 8915
chenal@msu.edu
www.phmtox.msu.edu/ & www.ns.msu.edu/neurosci/
Specialty Keywords: Vascular biology, Gene therapy, Oxidative
stress.

Dr. Alex Chen's research laboratory studies vascular biology and gene therapy, focusing on oxidative stress-induced vascular dysfunctions in hypertension, diabetes, and ischemic stroke. Conventional and confocal fluorescent microscopies are routinely used, among other techniques.

L.X. Li, E. Crockett, D.H. Wang, J.J. Galligan, G.D. Fink and A.F. Chen (2002). Gene transfer of endothelial NO synthase and manganese superoxide dismutase on arterial vascular cell adhesion molecule-1 expression and superoxide production in deoxycorticosterone acetate-salt hypertension. *Arterioscler. Thromb. Vasc. Biol.* **22**, 249-255.

Date submitted: 23rd August 2002

Herbert C. Cheung, Ph.D.

Department of Biochemistry and Molecular Genetics,
University of Alabama at Birmingham,
490 MCLM, 1918 University Boulevard,
1530 3rd Avenue South,
Birmingham, AL 35294-0005.

hccheung@uab.edu

Specialty Keywords: Motor proteins, Troponin, FRET.

My research is focused on the application of fluorescence in general, and FRET in particular, to mechanistic studies of motor proteins (muscle myosin and kinesis), molecular and structural aspects of calcium activation and regulation of cardiac myofilaments, modeling of the actomyosin cycle, complemented by collaborative efforts using molecular modeling and other forms of spectroscopy. Recently, we started FRET on proteins exchanged into single skinned muscle fibers for simultaneous correlation of conformational changes with fiber activation.
W-J. Dong, J. Xing, M. Villain, M. Hellinger, J. R. Robinson, M. Chandra, R. J. Solaro, P. K. Umeda, and H. C. Cheung (1999) *J. Biol. Chem.* **274**, 31382-31390.
W.-J. Dong, J. M. Robinson, J. Xing, P. K. Umeda, and H. C. Cheung (2000) *Protein Sci.* **9**, 280-289.

Date submitted: 4th September 2002   **Rivka Cohen-Luria, Ph.D.**

Chemistry, Ben-Gurion University,
P.O. Box 653, Beer Sheva,
Israel, 84105.
Tel: 972 8 6461191   Fax: 972 8 6472943
riky@bgumail.bgu.ac.il

Specialty Keywords: Prostaglandins, Membrane Dynamics, Lipid-Protein & Protein-Protein & Protein-Ligand / drug Interactions.

Research topics: the role of hydrophobic interactions in membranal and non-membranal protein function and regulation, signal transduction, cell cycle and proliferation, cell differentiation and intercellular interactions, angiogenesis, apoptosis, magnetic field effects on biological systems.
On the Regulatory Role of Dipeptidyl Peptidase IV (= CD26 = Adenosine Deaminase Complexing Protein) on Adenosine Deaminase activity. I. Ben-Shooshan, A. Kessel, N. Ben-Tal, R. Cohen-Luria and A.H. Parola .*Biochim. Biophys. Acta*, 1587, 21-30 (2002).
Nature of interaction between basic fibroblast growth factor and the antiangiogenic drug 7,7-(carbonyl-bis[imino-N-methyl-4,2-pyrrolecarbonylimino[N-methyl-4,2-pyrrole]-carbonylimino])-bis-(1,3-naphtalene disulfonate): 2. Removal of polar interactions affects protein folding. M. Zamai, C. Hariharan, D. Pines, M. Safran, A. Yayon, V.R. Caiolfa, R. Cohen-Luria, E. Pines and A.H. Parola. *Biophys. J.*, in press.

Date submitted: 12th September 2002   **Jeffrey J. Comerford, Ph.D.**

VARIAN

Life Science Group Product Manager,
Varian Australia Pty. Ltd,
679 Springvale Road,
Mulgrave, 3170, Australia.
Tel: +61 3 9566 1483   Fax: +61 3 9566 1196
jeff.comerford@varianinc.com
www.varianinc.com

Specialty Keywords: Life science, Molecular spectroscopy, Molecular biology, Analytical instrumentation.

My background is in molecular spectroscopy, in particular, the solution and photochemical behavior of square planar platinum(II) anti-cancer drugs. Experienced in the use of fluorescence, UV-Vis absorption and high pressure spectroscopy techniques with particular areas of interest including genomics, protein and cell based fluorescence applications, HTS assays and *ab initio* theoretical calculations. My current role is in marketing and business development, where I am responsible for Varian's fluorescence product line, which includes the Cary Eclipse fluorescence spectrophotometer.

Date submitted: 14th July 2002

Matthew Cook, Ph.D.

Acumen Bioscience Limited,
Melbourn Science Park, Cambridge Road,
Melbourn, Hertfordshire,
United Kingdom, SG8 6EE.
Tel: +44 (0) 1763 262233   Fax: +44 (0) 1763 266729
mcook@acumenbioscience.com
www.acumenbioscience.com

Specialty Keywords: Laser-Based Scanning, Fluorescent Detection, HTS.

Acumen Bioscience Ltd provides solutions to the drug discovery industry. The company develops and provides laser-based fluorescence detection instruments, assay protocols and reagents. The technologies combine high information screening with throughputs for both cell-based and cell-free assays.

The Acumen Explorer™ instrumentation uses of fluorescent dyes to monitor changes in intra and extra cellular biochemical events. The proprietary software algorithms allow measurements of cell morphology, size, and spectral characteristics utilizing either single or multiple fluorescent dyes.

Date submitted: 31st July 2002

Paulo J. G. Coutinho, Ph.D.

Departamento de Física, Universidade do Minho,
Campus de Gualtar,
4710-057 Braga,
Portugal.
Tel: + 351 253 604321 Fax: +351 253 678981
pcoutinho@fisica.uminho.pt
www.fisica.uminho.pt

Specialty Keywords: Kinetics in confined media, Biophysics, Nanoparticles production by surfactant templating.

Current interests: Biophysics, kinetics in confined media, self-assembly molecules, microaggregates (structure and applications), computer simulations, solar energy conversion, dynamics in biological membranes, photodegradation of pollutants, semiconductor nanoparticles, Langmuir-Blodgett films, surfactant templating.J. A. B. Ferreira, P. J. G. Coutinho, S. M. B. Costa, J. M. G. Martinho (2000), Dissociation Kinetics of Excited Rhodamine $^{3}B^{+}ClO_4^{-}$ in Water/toluene Mixtures: Dynamic Aspects, *Chem. Phys.* 262, 453.

A.L.F. Baptista, P.J.G. Coutinho, M.E.C.D. Real Oliveira, J.I.N. Rocha Gomes (2000), Effect of Surfactants in Soybean Lecithin Liposomes Studied by Energy Transfer between NBD-PE and N-Rh-PE, *J. Liposome Research*, 10, 419.

Date submitted: 28th August 2002

Scott D. Cummings, Ph.D.

Department of Chemistry, Kenyon College,
Gambier,
OH 43022,
USA.
Tel: 740 427 5355
cummingss@kenyon.edu
chem.kenyon.edu/faculty/cummings.htm

Specialty Keywords: Photoluminescent metal complexes.

Research with undergraduates at Kenyon College centers on the synthesis and spectroscopy of transition metal complexes having long-lived excited states. Special attention has focused on photoluminescent platinum (II) complexes capable of photo-induced electron transfer and energy transfer.

M. Cortes, J. D. Oppenheimer, K. E. Downey and S. D. Cummings (2002) "Photoinduced Electron Transfer and Energy Transfer Reactions of Hydroxo-(2,2':6',2"-terpyridine) Platinum (II)" *Inorganica Chimica Acta* **333**, 147-150.

S. E. Hobert, J. T. Carney, S. D. Cummings (2001) "Synthesis and Luminescence Properties of Platinum(II) Complexes of 4'-Chloro-2,2':6',2"-terpyridine and 4,4',4"-Trichloro-2,2':6',2"-terpyridine" *Inorganica Chimica Acta*, **318**, 89-96.

Date submitted: 22nd July 2002

Robert E. Dale, Ph.D.

University of London

The Randall Centre for Molecular Mechanisms of Cell Function, King's College London.
&
GKT School of Biomedical Sciences,
3rd Floor, New Hunt's House Guy's Hospital Campus,
London. SE1 1UL, UK.

Tel: 44 (0)207 848 6471   Fax: 44 (0)207 848 6435

Specialty Keywords: Orientation, Depolarization, FRET.

Theory and practice of steady-state and time-resolved fluorescence and fluorescence polarization spectroscopy and Förster long-range resonance excitation energy transfer (FRET) as probes of molecular, macro-molecular and supra-molecular structure and dynamics in their relation to biochemical and biological function and mechanism. Recent and current efforts centre on muscle cross-bridge orientation and dynamics by fluorescence depolarization, and location of Taxol™ binding site in microtubules by homogeneous FRET depolarization.

Date Submitted: 16th May 2002

Sabato (Tino) D'Auria, Ph.D.

Institute of Protein Biochemistry,
Via Pietro Castellino, 111,
Naples, 80131, Italy.
& C.F.S., Baltimore, MD, USA
Tel: +39 0816132250   Fax: +39 0816132270
dauria@dafne.ibpe.na.cnr.it / sabato@cfs.umbi.umd.edu

Specialty Keywords: Thermophilic enzymes, Fluorescence biosensor, Diabetes.

My primary research goal is to advance the field of protein research by investigating the functional and structural features of enzymes and proteins by biophysical approach. My research activity is focused on the study of the relationships of structure-function-stability, in enzymes and proteins. In particular, I am interested in the study of thermostable biomoleculesl isolated from thermophilic microorganisms. These macromolecules show an uncommon stability at high temperature, pressure, and in the presence of organic solvents and detergents. For this reason, since 1997 I am collaborating with the Center for Fluorescence Spectroscopy, Baltimore, USA for developing a new class of stable fluorescence protein sensors for sensing analytes of high clinical interest, such as glucose, sodium and lactate.

Date submitted: 13th September 2002

Lesley Davenport, Ph.D.

Department of Chemistry, Brooklyn College of CUNY,
2900 Bedford Avenue, Brooklyn,
New York 11210,
USA.
Tel: 718 951 5750   Fax: 718 951 4827
LDvnport@brooklyn.cuny.edu
academic.brooklyn.cuny.edu/chem/davenport/

Specialty Keywords: Time-resolved fluorescence, Lipid packing and dynamics, Fluorescent probes.

Research in our laboratory is currently focused on employing fluorescence methods for studying molecular interactions. We are particularly interested in employing long-lived fluorescence probes for investigating submicrosecond dynamics.

L. Davenport, B. Shen, T.W. Joseph and M.P. Straher (2001) A Novel Fluorescent Coronenyl-Phospholipid Analogue for Investigations of Submicrosecond Lipid Fluctuations. *Chem. Phys. Lipids*. **109**, 145-156.

P. Targowski and L.. Davenport (1998) Pressure Effects of Submicrosecond Phospholipid Dynamics Using a Long-Lived Fluorescence Probe, *J. Fluorescence*, **8**, 121-128.

Date submitted: 12th September 2002   **Rodrigo F. M. de Almeida (Ph.D. student)**

Centro de Química-Física Molecular, Instituto Superior,
Técnico, Av. Rovisco Pais
1049-001 Lisboa,
Portugal.
Tel: +351 218419248   Fax: +351 218464455
r.almeida@ist.utl.pt

Specialty Keywords: Membrane domains, Lipid bilayers, Lipid-protein interactions.

Phase separation in multicomponent lipid bilayers (domain formation and topology in binary and ternary membranes). Model systems for raft/non-raft coexistence Interaction of peptides with membranes and its relation with the phase behaviour/domain structure (mutual influence concerning extent of interaction, structure and dynamics).

R. F. M. de Almeida, L. M. S. Loura, A. Fedorov, and M. Prieto (2002) *Biophys. J.* **82**, 823-834.
L. M. Contreras, R. F. M. de Almeida, A. Fedorov, J. Villalaín, and M. Prieto (2001) *Biophys. J.* **80**, 2273-2283.

Date submitted: 3rd September 2002   **Frans C. De Schryver, Ph.D.**

Departmentof Chemistry, KULeuven,
Celestijnenelaan 200F Heverlee,
B-3001,
Belgium.
Tel: +3216327405   Fax: +3216327989
Frans.deschryver@chem.kuleuven.ac.be
www.chem.kuleuven.ac.be/research/mds/members.htm

Specialty Keywords: Time resolved fluorescence, Confocal microscopy, Single molecule spectroscopy.

The research group has over the years established an ensemble of techniques with special emphasis on pico second fluorescence decay acquisition and analysis by self developed algorithms (global and compartmental analysis), up-conversion and single molecule spectroscopy. The group has set up tools to down scale in size and in time the object of the photochemical and photophysical study.

M. Lor, R. De, S. Jordens, G. De Belder, G. Schweitzer, M. Cotlet, J. Hofkens, T. Weil, A. Herrmann, K. Müllen, M. Van der Auweraer, F.C. De Schryver J. Phys. Chem.,106, 10, 2083-2090 (2002) T. Vosch, J. Hofkens, M. Cotlet, F. Köhn, H. Fujiwara, R. Gronheid, K. Van Der Biest, T. Weil, A. Herrmann, K. Müllen, S. Mukamel, M. Van der Auweraer, F.C. De Schryver Angew. Chem., 40, , 4643-4648 (2001).

de Silva, A.P.
De, S.

Date submitted: 18th June 2002

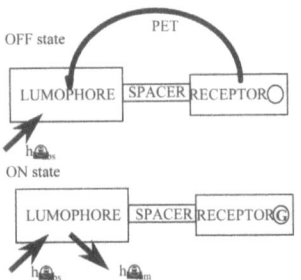

Amilra P, de Silva, Ph.D.

School of Chemistry, Queen's University,
Stranmillis Road, Belfast,
BT9 5AG,
Northern Ireland.
Tel: +44 28 90 27 44 22   Fax: +44 28 90 38 21 17
a.desilva@qub.ac.uk
www.ch.qub.ac.uk/staff/desilva/apds.html

Specialty Keywords: Luminescent sensors, Molecular logic.

We helped to establish the general principle of luminescent/fluorescent photoinduced electron transfer (PET) sensors which is now widely used. This principle is shown illustrated above where a guest G is being sensed by switching 'on' of luminescence. Such sensors can monitor protons, several metal ions, several anions as well as some larger molecular species. We also published the first work on molecular logic gates in the primary literature. This area also receives increasing attention since it permits molecules to process information with a growing complexity.

Date submitted: 15th August 2002

Soma De, Ph.D.

Laboratory of Molecular Biology and Biochemistry,
The Rockefeller University,
1230 York Avenue, New York City,
10021, NY, USA.
Tel: (212) 327 8284   Fax: (212) 327 7904
des@mail.rockefeller.edu

Specialty Keywords: Macular degeneration, Lipofuscin, Membrane Mimics.

Accumulation of lipofuscin in retinal pigment epithelial (RPE) cells constitute a predicament for the development of age-related macular degeneration. My research focuses on how A2E, a component of lipofuscin, causes the cellular membrane damage and induces the apoptosis of RPE cells. I have also applied fluorescence spectroscopy extensively to study the membrane properties of several synthetic dimeric lipid systems.

S. De and T. P. Sakmar, (2002). Interaction of A2E with model membranes. Implications to the pathogenesis of age-related macular degeneration. *J. Gen. Physiol.* **120**, 147-157. S. Bhattacharya and S. De, (1999). Synthesis and vesicle formation from dimeric pseudoglyceryl lipids with $(CH2)_m$ spacers: Pronounced m-Value dependence of thermal properties, vesicle fusion, and cholesterol complexation. *Chem. - A Eur. J.* **5**, 2335-47.

Date submitted: 27th August 2002

Todor G. Deligeorgiev, Ph.D.

University of Sofia, Faculty of Chemistry,
1, James Bourchier Avenue,
1126 Sofia,
Bulgaria.
Tel: +359 2 6256 269   Fax: +359 2 9625438
toddel@chem.uni-sofia.bg

Specialty Keywords: Dye synthesis, Fluorescence, Bioapplications of fluorescent probes.

In the last years our research were directed towards the synthesis of novel nucleic acid dyes based mainly on Thiazole Orange and Oxazole Yellow chromophores as non-covalent fluorescent probes. We are also interested in development of novel calcium probes based on coumarin fluorophores. Some analytical bioapplications of the novel probes were investigated too.

Liepouri F, Deligeorgiev TG, Veneti Z, Savakis C, Katerinopoulos HE, Near-membrane iminocoumarin-based low affinity fluorescent Ca^{2+} indicators, *Cell Calcium* **31**(5) (2002) 221-227.

Date submitted: 30th August 2002

James N. Demas, Ph.D.

Department of Chemistry, University of Virginia,
McCormick Rd., Charlottesville,
VA, 22904,
USA.
Tel: (434) 024 3343   Fax: (434) 924 3710
demas@virginia.edu
www.people.virginia.edu/~jnd

Specialty Keywords: Lifetime, Metal complexes, Sensors.

We are designing, synthesizing and characterizing highly emissive Os, Ir, Re, and Ru complexes for use as sensitizers, sensors (e.g., oxygen, CO_2), and molecular probes. We are also developing mathematical and instrumental methods for making measurements. Major concerns are the role of the polymer supports in the sensor properties, especially site heterogeneity. We are utilizing conventional fluorescence, confocal, and two photon microscopy to sort out these issues.

K. A. Kneas, *et al.* (2000). Fluorescence Microscopy Study of Heterogeneity in Polymer Supported, Luminescence Based Oxygen Sensors", Microscopy and Microanalysis, **6**, 551-561.

W. D. Bare, N. Mack, W. Xu, J. Demas, and B. DeGraff (2002). Multicomponent Lifetime-Based pH Sensors Utilizing Constant-Lifetime Probes, Anal. Chem., 74, 2198-2209.

Devaney, J.J.
Dobek, A.T.

Date submitted: 22nd August 2002

John J. Devaney

Boston Electronics Corporation,
91 Boylston Street, Brookline,
MA, 02445,
USA.
Tel: (800) 347 5445 or (617566 3821)   Fax: (617) 731 0935
fsp@boselec.com
www.boselec.com

Specialty Keywords: TCSPC, Spectroscopy, Photodetection.

Instrumentation Engineer at Boston Electronics Corporation, North American agents for Becker & Hickl GmbH of Berlin, Germany and for Edinburgh Instruments Ltd of Edinburgh, Scotland. Specialist in monochromators and spectrometers.

Date submitted: 27th August 2002

Andrzej T. Dobek, Ph.D.

Faculty of Physics, A.Mickiewicz University in Poznań,
Umultowska 85,
61-614 Poznań,
Poland.
Tel: +48 (61) 8295252   Fax: +48 (61) 8295155
dobek@amu.edu.pl
bio5.physd.amu.edu.pl

Specialty Keywords: Molecular biophysics, Photobiology, Ultra-fast laser spectroscopy.

Current Research Interests: Transient absorption, fluorescence and photovoltage studies of primary events in photosynthesis, static and dynamic light scattering in biomacromolecular solutions, nonlinear light scattering in solution of macromolecules oriented by DC magnetic field and optical field.

K.Gibasiewicz, R.Naskręcki, M.Ziółek, M.Lorenc, J.Karolczak, J.Kubicki, J.Goc, J.Miyake, and A.Dobek (2001). Electron transfer in the reaction center of the photosynthetic bacterium *Rh.sphaeroides* R-26 measured by transient absorption in the blue spectral range, *J.Fluorescence* **11**, 37-44 .

G.Paillotin, W.Leibl, J.Gapiñski, J.Breton and A.Dobek (1998). Light gradients in spherical photosynthetic vesicles, *Biophys.J.* **75**, 124-133 (1998).

Date submitted: 10ᵗʰ July 2002

Wen-Ji Dong, Ph.D.

Biochemistry and Molecular Genetics,
University of Alabama at Birmingham, MCLM 487,
1918 University Boulevard,
Birmingham, AL, 35294-0005, USA.
Tel: 205 934 2269   Fax: 205 975 4621
wdong@uab.edu

Specialty Keywords: Fluorescence, Kinetics, Cardiac thin filament protein.

The primary focus of my current research involves the application of fluorescence spectroscopy combining with molecular biology approaches in study of cardiac thin filament proteins and bioassay development, including study of calcium activation mechanism of cardiac muscle; elucidation of structure-function relationship within thin filament; and development and application of novel fluorescence and luminescence assay for biological studies and high throughput drug screening.

Dong *et al.* "Ca2+ induces an extended conformation of the inhibitory region of troponin I in cardiac muscle troponin." *J. Mol. Biol.* 314:51-61 (2001). 2.   Dong *et al.* "A kinetic model for the binding of Ca2+ to the regulatory site of troponin from cardiac muscle." *J. Biol. Chem.* 272:19229-19235 (1997).

Date submitted: 7ᵗʰ May 2002

Andrey O. Doroshenko, D.Sc., Ph.D.

Department of Physical Organic Chemistry,
Institute for Chemistry at
Kharkov V.N. Karazin National University,
4 Svobody sqr., Kharkov, 61077,
Ukraine.
Tel: +380572457335   Fax: +380572457130
andrey.o.doroshenko@univer.kharkov.ua

Specialty Keywords: High Stokes shift organic luminophores.
Design and investigation of abnormally high Stokes shift organic fluorescent species: sterically hindered aromatic heterocyclic molecules, excited state intramolecular proton transfer (ESIPT) compounds, cation-sensitive fluorescent probes, fluorescent probes for biomembrane studies. Elucidation of interrelations between the molecular structure and photophysical properties of organic compounds. Photochemical transformations of organic molecules. Quantum chemical modeling related to fluorescent and photochemical ability of organic luminophores.

Doroshenko A.O., Posokhov E.A., Verezubova A.A.,   Ptyagina L.M., Skripkina V.T., Shershukov V.M. 2002, Photochem. Photobiol. Sci., **1**, 92-99.
Doroshenko A.O., Baumer V.N., Verezubova A.A., Ptyagina L.M. 2002, J. Mol. Struct., **609**, 29-37.

Dougla, P.
Drexhage, K.H.

Date submitted: 9th August 2002   **Peter Douglas, Ph.D.**

Chemistry Department,
University of Wales Swansea,
Singlton Park, Swansea,
SA2 8PP, UK.
Tel: +44 (0) 1792 51308   Fax: +44 (0) 1792 295747
P.Douglas@swan.ac.uk

Specialty Keywords: Porphyrins, Optical sensors, Photographic dyes.
Photochemical research interests: photodegradation mechanisms of photographic and textile dyes, photochemistry on thin film TiO_2, luminescent oxygen sensors, thin film optical sensors for medical industrial and environmental applications, photochemistry of porphyrins and metalloporphyrins, colloidal photochemistry electrochemistry and reaction kinetics.

C.D.Geddes and P.Douglas, Fluorescent dyes bound to hydrophilic copolymers - applications for aqueous halide sensing, (2000), *App. Poly. Sci.,* **76**, 603-615.
P.Douglas and K.Eaton, Response characteristics of thin film oxygen sensors, Pt and Pd Octaethylporphyrins in polymer films, (2002) Sens. Actuators B, 200-208.

Date submitted: 10th July 2002   **Karl H. Drexhage, Ph.D.**

Chemistry Department, University of Siegen,
D-57068 Siegen,
Germany.
ATTO-TEC GmbH, D-57008 Siegen, Germany.
Tel: +49 (0) 271 740 4187   Fax: +49 (0) 271 7420502
drexhage@chemie.uni-siegen.de
drexhage@atto-tec.com

Specialty Keywords: Fluorescence, Organic Dyes, Fluorescent Labels.

Research Interests: My research is centered around the process of light emission by molecules and the influence of molecular structure on fluorescence. Research topics are: Inter- and intramolecular energy transfer, influence of a mirror on decay time and directional characteristics of fluorescence, cooling by anti-Stokes fluorescence, laser dyes, development of fluorescent labels for biochemistry and medicine.

J. Arden-Jacob, J. Frantzeskos, N.U. Kemnitzer, A. Zilles, and K.H. Drexhage (2001). New fluorescent markers for the red region, Spectrochim. Acta A, **57**(11), 2271-2283.

Date submitted: 30th August 2002

David T. F. Dryden, Ph.D.

School of Chemistry, University of Edinburgh,
The King's Buildings, Edinburgh,
EH 9 3JJ,
United Kingdom.
Tel: +44 131 650 4735
david.dryden@ed.ac.uk
www.chem.ed.ac.uk/staff/dryden.html & www.cosmic.ed.ac.uk/

Specialty Keywords: Protein-DNA interactions, Fluorescence spectroscopy, Single-molecule imaging.

I am interested in all aspects of protein and DNA structure and dynamics with particular emphasis on combining physical and biological techniques at the "interface" between the physical and life sciences.

M.D. Walkinshaw, P. Taylor, S.S. Sturrock, C. Atanasiu, T. Berge, R.M. Henderson, J.M. Edwardson, and D.T.F. Dryden. Structure of Ocr from Bacteriophage T7, a Protein that Mimics B-Form DNA. *Molecular Cell* [2002] **9**, 187–194.
The DNA binding characteristics of the trimeric *Eco*KI methyltransferase and its partially assembled dimeric form determined by fluorescence polarisation and DNA footprinting. L.M. Powell, B.A. Connolly & D.T.F. Dryden. [1998] *J. Mol. Biol.* **283**, 947-961.

Date submitted: 12th June 2002

Guy Duportail, Ph.D.

Laboratoire de Pharmacologie et Physicochimie,
Université Louis Pasteur, Strasbourg,
Illkirch, 67401,
France.
Tel: (33) 390 244 260   Fax: (33) 390 244 312
duportai@aspirine.u-strasbg.fr
umr7034.u-strasbg.fr

Specialty Keywords: Biophysics, Liposomes, Fluorescent Probes.

My major interest is for membrane biophysics by using fluorescence spectroscopy methods. Different fields of research are considered: membrane photophysics, development of liposomes as mimicking membrane systems, conception and study of new fluorescent probes for liposomes and biomembranes (DPH, pyrene and 3-hydroxyflavone derivatives), physicochemistry of the process of non-viral transfection by cationic lipids.
G. Duportail, and P. Lianos (1996) in *Vesicles,* Surfactant Science Series, Vol. 62 (M. Rosoff Ed.), Marcel Dekker, New York, pp. 295-372.
G. Duportail, A. Klymchenko, Y. Mély, and A. Demchenko (2001) *FEBS Letters* **508**, 196-200.

Dürkop, A.
Dyubko, T.S.

Date submitted: 9th September 2002   **Axel Dürkop, Ph.D.**

Institute of Analytical Chemistry, University of Regensburg,
Universitätsstraße 31, Regensburg,
93053,
Germany.
Tel: 0049 941 943 4053   Fax: 0049 941 943 4064
axel.duerkop@chemie.uni-regensburg.de

Specialty Keywords: Ruthenium Metal Ligand Complexes, Europium Complexes, Labels, Hydrogen Peroxide.

The research includes the synthesis of new luminescent ruthenium metal ligand complexes which can be conjugated to different functional groups in biolmolecules and the use of these complexes in various assay-formats (e.g. FRET-Immunoassays, DNA-Oligonucleotides). Further research is done about the use of luminescent europium complexes for the analysis of hydrogen peroxide in aqueous solutions as well as in enzymatic assays.

Polarization immunoassays using reactive ruthenium metal ligand complexes as labels, Dürkop, A., Lehmann, F., Wolfbeis, O. S. (2002), *Anal. Bioanal. Chemistry*, **372:** 688-694.

Europium Ion-Based Luminescent Sensing Probe for Hydrogen Peroxide, Wolfbeis, O. S., Dürkop, A., Wu, M., Lin Z. (2002), , *Angew. Chemie, accepted.*

Date submitted: 15th September 2002   **Tatyana S. Dyubko, Ph.D.**

Dept. of Cryobiophysics, Inst. for Problems of Cryobiology and Cryomedicine, Ukrainian National of Sciences,
23 Pereyaslavskaya Str., Kharkov,
61015, Ukraine.
Tel: + 380 572 7726141   Fax: + 380 572 7720084
cryo@online.kharkov.ua   &   (Inst.); dyubko@un.com.ua
www.geocities.com/dyubko2001/

Specialty Keywords: Fluorescence, Biophysics, Cryobiology.

The main directions of ones research is: (1) using fluorescent spectroscopy methods and its modifications, investigation of molecular mechanisms of cell membranes cryoinjury and cryoprotection; (2) development a fluorescent probes methods and its application to determination of human serum proteines and biological membranes structural rearrangements after non-physiological conditions action (low temperatures, ionizing radiation, non-organic solvents, etc.) and after some human diseases (ichemical heart diseases, tireotoxicosis, pregnancy toxicosis, jaundice et. al.); (3) testing of new fluorescent dyes with aim of its application in medical diagnostics; on the base of fluorescent approaches developed, search of new high sensitive and specific diagnostic methods of some human diseases. Author of more 90 scientific publications.

Date submitted: 9th August 2002   **Kay Eaton, Ph.D.**

Chemistry Department,
University of Wales Swansea,
Singlton Park, Swansea,
SA2 8PP, UK.
Tel: +44 (0) 1792 295506   Fax: +44 (0) 1792 295747
cmsolar@swan.ac.uk

Specialty Keywords: Optical oxygen sensors, Redox chemistry, Luminescence quenching.

Research interests: The development of novel luminescence and redox based optical oxygen sensors. Dye redox chemistry. Steady-state and time-resolved studies of metalloporphyrin luminescence quenching by oxygen. Kinetic modelling of oxygen quenching of luminescence in heterogeneous thin polymer films.

K. Eaton, A novel colorimetric oxygen sensor: dye redox chemistry in a thin polymer film, (2002), Sens. and Actuators B, **85**, 42-51.
P.Douglas and K.Eaton, Response characteristics of thin film oxygen sensors, Pt and Pd Octaethylporphyrins in polymer films, (2002) Sens. Actuators B, **82**, 200-208.

Date submitted: 9th July 2002   **Richard H. Ebright, Ph.D.**

Howard Hughes Medical Institute, Rutgers University,
190 Frelinghuysen Road,
Piscataway NJ 08902, USA.
Tel: 732 445 5179   Fax: 732 445 573
ebright@waksman.rutgers.edu
www.hhmi.org/research/investigators/ebright.html
waksman.rutgers.edu/Waks/Ebright/ebright.html
Specialty Keywords: Transcription, Labelling strategies, FRET, FP, Single moleculenanomanipulation, Single-molecule imaging.

Our group is interested in the first step in gene expression: i.e., transcription. Our objectives are: (i) to understand the structural and mechanistic basis of transcription initiation, (ii) to understand the structural and mechanistic basis of transcription activation, and (iii) to develop inhibitors of transcription initiation and transcription activation.

J. Mukhopadhyay, A. Kapanidis, V. Mekler, E. Kortkhonjia, Y. Ebright, and R. Ebright (2001) Translocation of σ^{70} with RNA polymerase during transcription: fluorescence resonance energy transfer assay for movement relative to DNA. *Cell* **106**, 453-463.
V. Mekler, E. Kortkhonjia, J. Mukhopadhyay, J. Knight, A. Revyakin, A. Kapanidis, W. Niu, Y. Ebright, R. Levy, and R. Ebright, R. (2002) Structural organization of RNA polymerase holoenzyme and the RNA polymerase-promoter open complex. *Cell* **108**, 599-614.

Date submitted: 1st September 2002

Hans-Joachim Egelhaaf, Ph.D.

Institute of Physical Chemistry, University of Tübingen,
Auf der Morgenstelle 8, D-72076 Tübingen,
Germany.
Tel: +49 7071 2976911   Fax: +49 7071 295490
hans-joachim.egelhaaf@ipc.uni-tuebingen.de
homepages.uni-tuebingen.de/hans-joachim.egelhaaf/

Specialty Keywords: Thin organic films, Molecular mobility, Fluorescence anisotropy.

Translational and rotational molecular mobilities in liquid-swollen polymers are investigated by steady-state and time-resolved fluorescence techniques (mainly quenching and anisotropy) in order to understand and control the accessibilities of polymer-bound active centers.

Photoinduced processes (e.g., charge carrier generation and recombination) in thin organic films of Pi-conjugated polymers are studied by steady-state and time-resolved absorption, fluorescence, and photoconductivity in order to elucidate the kinetics and mechanisms of these processes.

H.-J. Egelhaaf, D. Oelkrug, P. Herman, E. Holder, H.A. Mayer, E. Lindner (2001) *J. Mater. Chem.* **11**, 2445 – 2552.

G. Cerullo, G. Lanzani, S. deSilvestri, H.-J. Egelhaaf, L. Lüer, D. Oelkrug (2000) *Phys. Rev. B* **62**, 2429.

Date submitted: 12th August 2002

Alla V. Egorova, Ph.D.

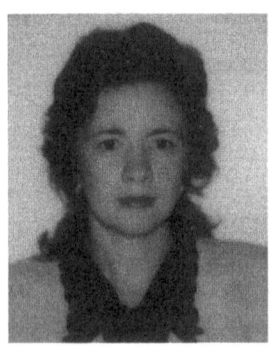

A.V.Bogatsky Physico-Chemical Institute,
of the National Academy of Sciences of Ukraine,
86 Lustdorfskaya doroga, Odessa,
65080, Ukraine.
Tel: + 38 (0482) 652 042   Fax: + 38 (0482) 652 012
physchem@paco.net

Specialty Keywords: Energy Transfer, Lanthanides, Fluorescent probes.

Investigation of intramolecular energy transfer from organic compounds to lanthanide ions. Optimization of conditions for the formation fluorescent complexes of organic compounds with lanthanide ions in solutions and on solid surfaces. Elucidation of interaction between molecular structure of organic ligands and fluorescent properties of investigated complexes. Design of fluorescent system for determination of drugs and fluorescent probes for fluoroimmunoassay.

A.Egorova, S.Beltyukova, O.Teslyuk, V.Karpinchik. J.Pharm. Biomed. Anal.,24 (2001) 1081-1085.

S.Beltyukova, O.Teslyuk, A.Egorova, E.Tselik. J. of Fluorescence, Vol.12, №2 (2002), 269-272.

Date submitted: 3rd September 2002

Benjamin Ehrenberg, Ph.D.

Department of Physics,
Bar Ilan University,
Ramat Gan, IL-52900,
Israel.
Tel: 00972 3 5353298   Fax: 00972 3 5353298
ehren@mail.biu.ac.il
www.ph.biu.ac.il/faculty/ehrenberg/

Specialty Keywords: Fluorescent probes, Photosensitizers, Porphyrins in membranes and cells.

We study the interactions of porphyrins and porphyrin-like molecules with artificial and natural membranes. The porphyrins are considered for use as photosensitizers for photodynamic therapy of malignancies and bacterial eradication. The aim of these studies is to understand the binding efficiency and topography of porphyrin sensitizers in membranes and to correlate these attributes with molecular structure. The extent of interaction, the depth of membrane-penetration and the efficiency of sensitized generation of singlet oxygen are monitored by fluorescence techniques.

Lavi A, Weitman H, Homes RT, Smith KM, Ehren berg B. 2002. The depth pf porphyrin in a membrane's physical properties strongly affect the photosensitizing efficiency. Biophys. J. 82:2101-2110.

Date submitted: 30th August 2002

Jörg Enderlein, Ph.D.

IBI-1, Forschungszentrum Jülich,
D-52425 Jülich,
Germany.
Tel: +49 2461 61 8069   Fax: +49 2461 61 4216
j.enderlein@fz-juelich.de
www.joerg-enderlein.de

Specialty Keywords: Single molecule fluorescence, Nanooptic.

Research centers on the fundamentals and applications of single molecule fluorescence spectroscopy (SMS). Special areas of expertise are fluorescence fluctuation spectroscopy, time-correlated single photon counting, statistics of SMS, wave-optical modeling of SMS experiments (Ref.1). Recently, the interaction of single fluorophores with metallic nano-cavities and complex nano-structures has become a central topic of his research.

C. Zander, J. Enderlein, R.A. Keller (Eds.) "Single-Molecule Detection in Solution - Methods and Applications" (Wiley-VCH, Berlin, 2002).

Date submitted: 4th July 2002

Yves Engelborghs, Ph.D.

Laboratory of Biomolecular Dynamics,
Katholieke Universiteit Leuven,
Celestijnenlaan 200 D, Leuven,
Belgium, B3001.
Tel: +32 16 32 71 60   Fax: +32 16 32 79 74
Yves.Engelborghs@fys.kuleuven.ac.be
www.chem.kuleuven.ac.be/research/bio/webye_en.html

Specialty Keywords: Proteins, Tryptophan, Lifetime, Correlation, Confocal, Phase.

Protein dynamics is the main interest. For many proteins single-W mutants were constructed and analyzed by phase fluorimetry and the fluorescence properties related to the environment of the W, and the rotameric state. Conformational changes are studied by kinetic techniques and modeled by "Targeted Molecular Dynamics". Fluorescence Correlation and Cross Correlation and Confocal scanning are used to study protein-protein and protein-nucleic acid and protein-drug interactions in vitro and in the living cell.

Y. Engelborghs (2001) Spectrochimica Acta Part A 57, 2255-2270.
E. Van Craenenbroeck et al. (2001) Biol. Chem. 382, 355-361.

Date submitted: 14th August 2002

Richard M. Epand, Ph.D.

Department of Biochemistry,
McMaster University,
1200 Main Street West,
Hamilton, ON L8N 3Z5, Canada.
Tel: 905 525 9140 ext: 22073
epand@mcmaster.ca

Specialty Keywords: Membrane, Hydrophobic, Liposomes.

We are interested in the use of fluorescence to determine membrane properties. We have studied the application of fluorescent probes for monitoring the nature of the membrane interface (1) and have also used fluorescence to identify protein sites that would facilitate membrane binding (2).
R.F. Epand, R. Kraayenhof, G.J. Sterk, H.W. Wong Fong Sang, and R.M. Epand (2002). Flourescent probes of membrane surface properties. *Biochem.Biophys.Acta* **1284,** 191-195.
D.L. LeDuc, Y.K. Shin, R.F. Epand, and R.M. Epand (2000). Factors determining vesicular lipid mixing induced by shortened constructs of influenza hemagglutinin. *Biochemistry* **39,** 2733-2739.

Date submitted: 3rd June 2002

Rainer Erdmann, Dipl.-Phys.

PicoQuant GmbH,
PicoQuant GmbH Rudower Chaussee 29 (IGZ),
12489 Berlin, Germany.
Tel: +49 30 6392 6560   Fax: +49 30 6392 6561
erdmann@pq.fta-berlin.de
www.picoquant.com

Specialty Keywords: Time correlated single photon counting (TCPC), Single molecule detection, Picosecond diode lasers.

We focus our R/D at PicoQuant on ultrasensitive fluorescence detection methods. Beside the development of components (like compact picosecond diode lasers, PC boards for TCPC, PMT detector modules) we design complete fluorescence spectrometers for various applications including comprehensive data analysis software. Furthermore we develope microscope based systems for fluorescence lifetime imaging (FLIM) applications. These systems offer ultimate sensitivity as well as highest spatial resolution as needed for single molecule detection. Beside traditional fluorescence correlation and fluorescence lifetime analysis we work on a combination of both methods.

Böhmer M., Wahl M., Rahn H.-J., Erdmann R., Enderlein J., Time-resolved fluorescence correlation spectroscopy, Chemical Physics Letters, Vol.353, No.5-6, pp.439-445 (2002).

Date submitted: 2nd September 2002

Anna M. Eremenko, Ph.D.

Natrional Ukrainian Academy of Sciences,
Institute of Surface Chemistry,
17 General Naumov str,
03164 Kiev, Ukraine.
Tel: 380 44 4449698   Fax: 380 44 4443567
annerem@mail.kar.net

Specialty Keywords: Charge transfer, Silica, Photocatalysis.

The scope of scientific interests concerns the problems of surface photochemistry. Fluorescent properties of adsorbed polyacenes on silica, silica-alumina and silica-titania surfaces: processes of intermolecular charge transfer, decay of fluorescence, formation of excimer, exciplex and CTC on the surfaces. Effect of surface active centers on the intramolecular charge transfer and conformational mobility of adsorbed TICT molecules. Luminescent diagnostic of active centers of silica, and mixed oxides. Sensibilization of titania photocatalysts to the visible with adsorbed excited organic dyes.

A. Eremenko, N.Smirnova, O.Rusina, O.Linnik, L.Spanhel, K.Rechthaler, Photophysical properties of organic fluorescent probes on nanosized TiO_2/SiO_2 systems J.Mol. Struct. 2000, 553/1-3, 1.

Erker, W.
Erostyák, J.

Date submitted: 31st July 2002   **Wolfgang Erker, Ph.D.**

Institute for Molecular Biophysics,
University of Mainz,
Welderweg 26, 55128 Mainz,
Germany.
Tel: 06131 39235 68   Fax: 06131 39235 57
wolfgang@biophysik.biologie.uni-mainz.de
biophys.biologie.uni-mainz.de/

Specialty Keywords: Proteins, Hemocyanin, Tryptophans.

My research is focused on the structure-function-relationship of proteins particularly hemocyanins. I am looking for conformational changes, investigating the flexibility and stability of the proteins. This involves energy transfer calculations, ensemble and single-molecule measurements especially with the intrinsic fluorophor tryptophan.

Lippitz M, Erker W, Decker H, van Holde KE, Basché T: Two-photon excitation microscopy of tryptophan containing proteins; Proc. Nat. Acad. Sci. 2002, 99 (5), 2772-2777.
Erker W, Hübler R, Decker H: Multi-donor- and multi-acceptor-quenching of oxy-hemocyanins by Förster transfer; Protein Science 2001, 10 (Suppl. 1), 144.

Date submitted: 23rd August 2002   **János Erostyák, Ph.D.**

Department of Experimental Physics, University of Pécs,
Ifjúság u. 6., Pécs,
H-7624,
Hungary.

erostyak@fizika.ttk.pte.hu
physics.ttk.pte.hu/erostyak/

Specialty Keywords: Dielectric relaxation, Energy transfer.

Present research interests are: Dielectric relaxation of dyes and proteins; Intra- and intermolecular energy transfer; Computer modelling of excited state processes; Analytical applications of dye-trace detection.
Experimental practice: phase fluorometry, femtospectrometry, laser fluorometry.

Date submitted: 6th September 2002

Kadriye Ertekin, Ph.D.

University of Dokuz Eylul,
Izmir,
Buca,
Turkey.
Tel: 090 232 388 8264
kadriye.yusuf@superonline.com
ertekin@sci.ege.edu.tr

Specialty Keywords: Optic sensor, Optical pH sensing, Carbon dioxide oxygen and Cation sensing.

My research covers investigation of pH, carbon dioxide, oxygen or cation (Na^+,K^+,Zn^{++}), sensitive fluorescent molecules in solid matrices. I also determine the photophysical and photochemical properties of such molecules both in conventional solvents and solid matrices (Sol-gel,Ethyl cellulose, PVC) by means of spectroscopic techniquies.

Ertekin, K., Alp, S., Karapire,C., Yenigul, B., Henden, E., Icli, S., Fluorescence emission studies of an azlactone derivative in polymer films; An optical sensor for pH measurements, Journal of Photochemistry and Photobiology A: Chemistry 137 (2000)155-161.
K.Ertekin, B.Yenigül, E.U.Akkaya, "Poassium sensing by using a newly synthesized squaraine dye in sol-gel matrix", Journal of Fluorescence, Vol 12 No 2, 2002, 263-268.

Date submitted: 29th August 2002

Jose Paulo S. Farinha, Ph.D.

Centro de Quimica-Fisica Molecular,
Instituto Superior Tecnico,
Av. Rovisco Pais, Lisboa, 1049-001,
Portugal.
Tel: 351 218419221   Fax: 351 218464455
farinha@ist.utl.pt
dequim.ist.utl.pt/docentes/3296

Specialty Keywords: Fluorescence Energy Transfer, Polymers, Colloids.

Study of polymer and colloidal systems using fluorescence techniques. Use of excimer formation to study the dynamics of polymer chains in solution. Study of the interface structure, colloidal particles (latex, micelles, etc) and in polymer films using non-radiative energy transfer [Ref 1,2]. Synthesis and dye-labeling of polymers. Modeling of the energy transfer kinetics in dispersed colloidal particles, polymer blend films, and other structured materials. Modeling of the diffusion in dye-labeled latex films. Static and dynamic (ps resolution) fluorescence measurements.

Farinha, J. P. S. et al *J. Phys. Chem. B* **1999**, *103*, 2487.
Farinha, J. P. S. et al *J. Phys. Chem.* **1996**, *100*, 12552.

Date submitted: 4th July 2002

Suren S. Felekyan, Ph.D.

Max-Planck-Institut,
Fur Biophysikalische Chemie,
Karl-Friedrich-Bonhoeffer-Institut,
abt. Spektroskopie und Photochemische Kinetik,
Am Faßberg 11, Göttingen D-37077, Germany.
Tel.: +49 551 201 1087   Fax: +49 551 201 1006
sfeleky@gwdg.de
www.mpibpc.gwdg.de/abteilungen/010/seidel/

Specialty Keywords: FCS, BIFL, FRET.

"Multidimensional single molecule fluorescence spectroscopy of biomolecules:
Screening applications and time-resolved investigation on biological processes".

Date submitted: 23rd August 2002

Maria L. Ferrer, Ph.D.

Instituto de Ciencia de Materiales,
Consejo Superior de Investigaciones Científicas,
Campus de Cantoblanco,
Madrid, 28049, Spain.
Tel: 34 91 334 90 00
mferrer@icmm.csic.es
www.icmm.csic.es/solgel/

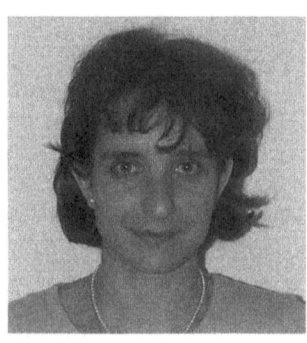

Specialty Keywords: Sol-Gel, Bioencapsulation, Fluorescence
sensing.

My research is focused on the preparation of organically modified silicates (Ormosils) through
the sol-gel method for optical applications. I have studied the chemical properties of the porous
surface of Ormosils through fluorescence spectroscopy. More recently, I am interested on the
encapsulation of biomolecules in sol-gel matrices and on the study of the structural integrity,
activity and fluorescence sensing applications of the encapsulated biomolecules.

M.L. Ferrer, F. del Monte and D. Levy (2001) Microviscosities at the Porous Cage of Silica
Glasses and Ormosils through Fluorescence Anisotropy J. Phys. Chem. B 105 (45) 11076-11080.
M.L. Ferrer, J. Gomez, C. R. Mateo, F. del Monte and D. Levy (2003) Denaturation Studies of
Horseradish Peroxidase Encapsulated in Sol-Gel Matrices. J. Sol-Gel Sci. & Tech. (in press).

Date submitted: 29th August 2002

Vlastimil Fidler, Ph.D.

Department of Physical Electronics, Faculty of Nuclear
Sciences and Physical Engineering,
Czech Technical University in Prague, V Holesovickach 2
Prague 8, Czech Republic, CZ – 180 00,
Tel: +420 22191 2221   Fax: +420 28468 4818
fidler@troja.fjfi.cvut.cz,
Vlastimil_Fidler@brown.edu
Affiliated: Brown University, Providence, R.I. 02912, USA.
Specialty Keywords: TR Fluorescence, Excitation Energy
Transfer, Photo-induced Intramolecular Processes.

Graduated from Charles University in Prague, long-term stays at the Royal Institu-tion, London,
UK, at IMS Okazaki, Japan, and at University of Chicago, USA; currently with CTU in Prague,
Czech Republic & affiliated with Brown University, Providence, USA.
Current topics of primary interest: Ultrafast TR fluorescence, kinetics and anisotropy; Intra-
molecular energy &electron transfer and re-distribution; Photo-physics of molecular switching.

V. Fidler, P. Kapusta, M. Nepras, J. Schroeder, I.V. Rubtsov, and K. Yoshihara (2002).
Femtosecond Fluorescence Anisotropy Kinetics as a Signature of Ultrafast Electronic Energy
Transfer in Bichromophoric Molecules: *Z. Phys. Chem,*. **216**, 589-603.

Date submitted: 12th September 2002

Judit Fidy, D.Sc., Ph.D.

Dept. of Biophysics and Radiation Biology,
Semmelweis University,
Puskin u. 9, Budapest,
H-1088, Hungary.
Tel: 36 1 266 2755 / 4052   Fax: 36 1 266 6656
judit@puskin.sote.hu

Specialty Keywords: Protein dynamics, Aggregation, Folding.

Research interests: Prof. Fidy's interest in protein dynamics started by detailed fluorescence
line narrowing studies on hemoproteins. On this basis she initialized a collaboration with Prof.
J. Friedrich (Bayreuth, D.) to perform the first spectral hole burning studies under high
pressure on a protein. Since 1993 she has her own research lab.in Budapest equipped with
FLN, various luminescence methods, cryostats, high pressure cells and computer capacity for
molecular modeling. They study the connection between protein dynamics and functionality.

J. Fidy et al., invited review, BBA, (1998) **1386,** 289-303.
L. Smeller , J. Fidy, Biophys.J. (2002) **82,** 426-436.

Fischer, P.
Fisz, J.J.

Date submitted: 28th August 2002

Peter Fischer

JenLab Gmbh,
Winzerlaer Strasse 2a,
Germany.
Tel: +49 3641 508220   Fax: +49 3641 508617
info@jenlab.de
www.jenlab.de

Specialty Keywords: Multiphoton imaging, FRET, Fluorescence lifetime imaging, Optical biopsy, Drug screening.

R/D is focussed on femtosecond laser systems for biotechnology, cell biology and medicine. Products include the multiphoton fluorescence imaging system *DermaInspect 100* for skin diagnostics and drug screening and the scanning microscope *TauMap* for fluorescence lifetime imaging including time-resolved FRET. In addition, miniaturized low-cost sterile cell chambers for long-term fluorescence microscopy and GFP imaging (*MiniCeM*) and fluorescent markers *JenFluor* for enzyme detection (e.g. alkaline phosphatase) are produced. Current development includes systems for nanosurgery with sub-200nm-cut sizes combined with imaging.

König et al. Optics Express 10(2002)171-176.
König et al.: SPIE Proceed 4620(2002)191-201.

Date submitted: 13th September 2002

Jacek J. Fisz, Ph.D.

Optical Spectroscopy and Molecular Engineering Group,
Institute of Physics, N. Copernicus University,
ul. Grudziadzka 5 / 7,
PL 87-100 Torun, Poland.
Tel: +48 56 6113297   Fax: +48 56 6225397
jjfisz@phys.uni.torun.pl

Specialty Keywords: Molecular fluorescence, Photochemistry, Photovoltaic systems.

Research fields*:* one- and two-photon excitation spectroscopy of solutions and organized media, evanescent wave excitation fluorescence and second-harmonic generation on organized molecular assemblies, excited-state processes in solutions and organized media, structural and dynamic properties of ordered molecular media, time-resolved fluorescence spectroscopy with polarized light.

J.J. Fisz, M.P. Budzinski, Fluorescence depolarization in organized media. Two-excited-state reactions controlled by orientation-dependent kinetic rates. I. Theory, J. Chem. Phys. 115 (15) (2001) 7130-7143.

J.J. Fisz, A method for visual and numerical recovery of state-dependent character of fluorophore-matrix aligning interactions, Chem. Phys. Letters 355 (2002) 94-100.

Date submitted: 17th June 2002

Danuta Frąckowiak, (Jabłoński) Ph.D.

Poznan University of Technology,
Institute of Physics,
Nieszawska 13A, 60 965 Poznań,
Poland .
Tel: +48 (61) 665 3180   Fax: +48 (61) 665 3201
frackow@phys.put.poznan.
www.put.poznan.pl

Specialty Keywords: Polarized light spectroscopy.
Current Research: 1) Investigations of the fate of absorbed energy in photosynthetic organisms, in their parts and in their anisotropic models by the measurements of polarized light fluorescence, delayed fluorescence and steady state photoacoustic spectra. The evaluation of the yield of triplet states generation using laser induced optoacoustic spectroscopy. 2)The measurements of the fluorescence of various dye-photosensitizers in healthy and cancerous cells as well as of the endogenous emission of stained cell material are due in order to select dyes suitable for photodynamic therapy and photodynamic diagnosis of cancer. From emission of irradiated stained cells the courses of photoreactions are established.

D.J. Qian, A. Planner, J. Miyake, D. Frąckowiak (2001). Photothermal effects and fluorescence spectra of tetrapyridylporphyrins, *J. Photochem. Photobiol. A: Chemistry*, **144**, 93-99

Date submitted: 30th August 2002

Gerasim Stoychev Galitonov, Ph.D.

Dept of Biophysics, Inst. of Exptl Physics,
University of Warsaw,
Zwirki i Wigury 93, PL-02089 Warsaw,
Poland.
Tel: +(004822) 5540715   Fax: +(004822) 5540001
Gierasim@yahoo.com
www.fuw.edu.pl

Specialty Keywords: Fluorescence, Enzyme-ligand interactions, State-of-the-art equipment.

Some of my interests are: Ligand tautomeric form identification and charge distribution in enzyme complexes by steady-state quenching. Rotamer identification in enzyme complexes by FRET, time-resolved and anisotropy measurements. Fluorescence and phosphorescence art-of-the-state equipment. Analysis of human genome sequences.

Stoychev G., Kiedaszuk B. & Shugar D. (2001) Interaction of *E. coli* PNP with the cationic and zwitterionic forms of the fluorescent substrate m^{7}Guo, *BBA*, **1544** (1-2) 74-88.
Stoychev G., Kiedaszuk B. & Shugar D. (2002) Xanthosine and xanthine: Substrate properties with PNP, and relevance to other enzyme systems, *Eur J Biochem*, **269** (16) 4048-4057.

Date submitted: 29th April 2002

Ashok Ganesan, M.Sc.

Department of Physics and Applied Physics,
University of Strathclyde,
107 Rottenrow,
Glasgow, G4 0NG,
Scotland, UK.
Tel: +44 (0)141 548 3059   Fax: +44 (0)141 552 2891
ashok.ganesan@strath.ac.uk

Specialty Keywords: Multiphoton, Melanin, Urocanic acid.

My research interest includes multiphoton induced fluorescence studies of skin chromophores. Areas of study encompasses: one- and multiphoton excited, time resolved fluorescence spectroscopy of melanin and urocanic acid isomers.

Date submitted: 16th August 2002

Fang Gao, Ph.D.

Department of Chemistry,
University of Tennessee,
Knoxville, TN 37919-1600,
USA.
Tel: 865 974 3473   Fax: 865 974 3454
fgao1@utk.edu

Specialty Keywords: Organic synthesis, Polymer & photo-polymer, Photochemistry.

Dr. Fang Gao's research focused on the design and synthesis novel dye molecules and photochemistry. He successfully dealt with a long-standing problem, i.e. if we can obtain the high efficiency of long-wavelength photo-polymerization systems. Dr. Gao synthesized many novel dye molecules, including coumarin dyes and cyanine dyes, and he found that these compounds all can be efficiently used as mono or two-photos sensitizers for photo-polymerization, and he illustrated the dye-sensitized photo-polymerization mechanism by various experimental method. He further studied the photoimaging of various dye-sensitized photo-polymerization systems. Dr. Gao also employed fluorescence probe to determine the formation of micelles and vesicles. His current research is involved with the preparation of new chiral molecules (life hand or right hand) and their photochemistry. Dr. Gao authored or coauthored 25 papers on dye and chiral molecules, photochemistry and photo-polymerization.

Date submitted: 9th August 2002

Michael S. Garley, Ph.D.

Chemistry Department,
University of Wales Swansea,
Singlton Park, Swansea,
SA2 8PP, UK.
Tel: +44 (0)1792 295796 Fax: +44 (0)1792 295747
M.S.Garley@swan.ac.uk

Specialty Keywords: Computer modelling, Chemical kinetics.

research interests: chemical kinetics, computer modelling, time resolved fluorescence and phosphorescence.

R.J.Berry, P.Douglas, M.S.Garley, D.Clarke, C.Winscom, Triplet energies, singlet-state properties and singlet oxygen quenching rate constants and quantum yields for two cyan azamethine dyes (1999), *J.Photochem.Photobiol.A.*, **120**, 29-36.
H.N.McMurray, P.Douglas, C.Busa and M.S.Garley, Oxygen quenching of tris(2,2'-bipyridine) ruthenium (II) in thin organic films, (1994) *J.Photochem.Photobiol.A.*, **80**, 283-288.

Date submitted: 13th September 2002

Sergiy V. Gatash, Ph.D.

Department Biological and Medicine Physics,
School of Radiophysics,
V.N. Karazin Kharkov National University,
4 Svobody Sq. Kharkov, 61077, Ukraine.
Tel: (38) 0572 45 72 12 Fax: (38) 0572 35 39 77
Sergiy.V.Gatash@univer.kharkov.ua

Specialty Keywords: Fluorescence spectroscopy, Hydrophobic and hydrophilic fluorescence probes.

Current Research Interests: My research is focused on investigation by means of fluorescence probes the conformation transitions of protein macromolecules especially fibrinogen and serum albumin.   I also study the influence of physical factors such as temperature and irradiation on conformation and function macromolecules and biological membranes.

Gatash et al., Influence of irradiation and low temperatures on structure-dynamical state of blood proteins. // Biophysical Bulletin, Issue 2 (11), (Visn. Khark. univ.)-2002.- p.46-49.
Andreeva et al., Influence of freezing on spectral properties of fibrinogen solutions. //Problems of Cryobiology, 1998, No 3, pp. 18-21.

Date submitted: 7th July 2002

Ehud Gazit, Ph.D.

Molecular Microbiology and Biotechnology,
Tel-Aviv University,
Tel-Aviv 69978,
Israel.
Tel: +972 3 640 9030   Fax: +972 3 640 5448
ehudg@post.tau.ac.il
www.tau.ac.il/lifesci/biotechnology/gazit/gazit.htm

Specialty Keywords: Protein Folding, Unfolding, and Misfolding.

In our lab we study protein folding, unfolding and misfolding using a variety of biochemical and biophysical techniques. A partial list of the experimental systems includes several bacterial toxin-antidote systems, type II diabetes-related amyloidogenic peptides, and the VHL tumor suppressor protein. Another line of research is directed toward the elucidation of the mechanism of "chemical chaperons" activity and their effect on folding, aggregation and amyloid formation.

Gazit, E. (2002) *Angew. Chem. Int. Ed.* 114, 257-259.

Gazit, E. (2002) *FASEB J.* 16, 77-83.

Date submitted: 5th May 2002

Chris D. Geddes, Ph.D.

Center for Fluorescence Spectroscopy,
Medical Biotechnology Center,
University of Maryland, 725 West Lombard St,
Baltimore, Maryland, 21201, USA.
Tel: 410 706 3149   Fax: 410 706 8409
Chris@cfs.umbi.umd.edu
cfs.umbi.umd.edu

Specialty Keywords: Principles of Fluorescence, Fluorescence Sensing, Metal-Enhanced Fluorescence.

Current Research Interests: The interactions of metallic surfaces with fluorophores, recently termed Radiative Decay Engineering and also Metal-enhanced Fluorescence. I am particularly interested in how modifications of radiative decay rates and/or excitation rates can effect fluorescence phenomenon, such as RET and MPE fluorescence, and the subsequent biomedical applications thereof. I am also interested in fluorescence sensing and multi-photo microscopy.

C. D. Geddes, J. Karolin and D. J. S. Birch (2002). 1 and 2-photon fluorescence anisotropy decay in silicon alkoxide sol-gels: *J. Phys. Chem. B*, **106**(15), 3835-3841.

Chris D .Geddes (2001). Optical Halide sensing using fluorescence quenching: Theory, Simulations and Applications, *Meas. Sci. Technol.*, **12**(9), R53-R88, 2001.

Date submitted: 26th August 2002

Hans C. Gerritsen, Ph.D.

Molecular Biophysics, Debye Institute,
Princetonplein 1, Utrecht,
NL-3584 CC,
Netherlands.
Tel: +31 30 2532824   Fax: +31 302532706
H.C.Gerritsen@phys.uu.nl
www1.phys.uu.nl/wwwmbf/

Specialty Keywords: FLIM, SPIM, CLSM, TPE.

Main areas of research are the development and application of new methodologies in fluorescence microscopy. This includes Fluorescence Lifetime Imaging, Spectral Imaging, FRET imaging, Single Molecule Imaging and Multi-Photon Excitation imaging. In addition work is carried out on the characterization of fluorescent probes and novel fluorescent markers such as quantum dots and fluorescent colloids. Applications include live cell imaging, ion concentration imaging and FRET based co-localization studies.

Quantitative pH imaging in cells using confocal fluorescence lifetime imaging microscopy. R. Sanders et al. (1995) Anal. Biochem., **227**, 302-308.

Photooxidation and photobleaching of single CdSe/ZnS quantum dots probed by room-temperature time-resolved spectroscopy. van Sark et al. (2001) J.Phys.Chem. B **105**, 8281-8284.

Date submitted: 13th May 2002

Ken P. Ghiggino, Ph.D.

School of Chemistry,
University of Melbourne,
Victoria, 3010,
Australia.
Tel: +61 (3) 8344 7137   Fax: +61 (3) 9347 5180
ghiggino@unimelb.edu.au
www.chemistry.unimelb.edu.au

Specialty Keywords: Ultrafast spectroscopy, Fluorescence imaging, Polymer photophysics.

Current interests: Studies of energy and electron transfer in multichromophoric assemblies using ultrafast spectroscopy techniques. Relaxation dynamics and energy migration in macromolecules studied by time-resolved fluorescence anisotropy measurements. Photophysics and time-resolved fluorescence imaging of photosensitizers for phototherapy.

T.A. Smith, D.J.Haines and K.P. Ghiggino (2000) Steady-state and time-resolved fluorescence polarization behaviour of acenaphthene, *J. Fluorescence* **10**, 365-373.

E.K.L. Yeow, K.P. Ghiggino, J.N.H. Reek, M.J. Crossley, A.W. Bosman, P.H. Schenning and E.W. Meier (2000) The dynamics of electronic energy transfer in novel multiporphyrin functionalized dendrimers: A time-resolved fluorescence anisotropy study, *J. Phys. Chem. B* **104**, 2596–2606.

Date submitted: 23rd July 2002

Cees Gooijer, Ph.D.

Analytical Chemistry & Applied Spectroscopy, Laser Centre,
Vrije Universiteit Amsterdam,
de Boelelaan 1083, Amsterdam,
1081 HV,The Netherlands.
Tel: +31 20 4447540   Fax: +31 20 4447543
gooijer@chem.vu.nl
www.chem.vu.nl/acas/

Specialty Keywords: High-resolution molecular fluorescence, Phosphorescence detection in LC and CE, Time-resolved laser fluorescence, Temperature jump.

Applied spectroscopy research is conducted in the Laser Centre Vrije Universiteit along an analytical chemistry line – in close cooperation with chromatographers with emphasis on hyphenated techniques – and a physical chemistry line focusing on the dynamics of the interaction between small molecules and (bio)macromolecules. Research topics are hyphenation of Raman spectroscopy and LC/CE; phosphorescence detection in CE; laser fluorescence detection including FRET; temperature jump/time-resolved fluorescence and cryogenic high-resolution molecular fluorescence.

Kuijt, J., Teylingen, R. van, Nijbacker, T., Ariese, F., Brinkman, U.A.T. & Gooijer, C. (2001). Detection of nonderivatized peptides in capillary electrophoresis using quenched phosphorescence. Analytical Chemistry, 73, 5026-5029.

Date submitted: 7th July 2002

Karl O. Greulich, Ph.D.

Inst.Mol.Biotech,
Postfach 100 813 Jena,
D 07708,
Germany.
Tel: +49 3641 656400   Fax: +49 3641 656410
kog@imb-jena.de
www.imb-jena.de/greulich

Specialty Keywords: Optical tweezers, Single molecules, Comet assay.

Current Interests: Reactions of single enzyme molecules are studied, for example the conversion of fluorescing. NADH into dark NAD+ by lactate dehydrogenase and the sequence specific cutting of fluorescently labeled individual DNA molecules, held by optical tweezers, with restriction endonucleases. Also, the fragility of genomes and genome regions of individual cells is visualized with the fluorescent COMET assay and COMET FISH.

K.O.Greulich 1999 Birkhäuser Basel Wien Boston (Monography) Micromanipulation by light in biology and medicine: The laser microbeam and optical tweezers.

B.Schäfer, H. Gemeinhardt and K.O.Greulich 2001 Angew.Chem.Int.Ed.4663-4666 Direct microscopic observation of the time course of single molecule DNA restriction reactions.

Date submitted: 12th September 2002

Ulrich-W. Grummt, Ph.D.

Friedrich-Schiller-Universitaet Jena,
Institute of Physical Chemistry,
Helmholtzweg 4, Jena,
Germany, D 07743.
Tel: +49 3641 948350   Fax: +49 3641 948302
cug@uni-jena.de
www.uni-jena.de/chemie/institute/pc/grummt

Specialty Keywords: Time correlated single photon counting with ps and ns time resolution, Polymer Photophysics.

Main research topic is photophysical chemistry of conjugated, luminescent polymers and functionalized dyes with potential application in solar energy conversion, molecular electronics, non-linear optics, optical information recording, and chemical sensing. Energy migration and electron transport are of particular interest.

Ab-initio and DFT quantum chemical calculations are used to support and interpret experimental results.

E. Birckner, U.-W. Grummt, A. H. Göller, T. Pautzsch, D. A. M. Egbe, M. Al-Higari, and E. Klemm, J. Phys. Chem. A 105 (2001) 10307 – 10315.

U.-W. Grummt, E. Birckner, M. Al-Higari, D. A. M. Egbe, and E. Klemm, J. Fluoresc. 11 (2001) 41 – 51.

Date submitted: 29th July 2002

Ignacy Gryczynski, Ph.D.

Center for Fluorescence Spectroscopy,
Medical Biotechnology Center,
University of Maryland, 725 West Lombard St,
Baltimore, Maryland, 21201, USA.
Tel: 410 706 7500   Fax: 410 706 8408
ignacy@cfs.umbi.umd.edu
cfs.umbi.umd.edu

Specialty Keywords: Fluorescence, Spectroscopy, Polarization, Multi-Photon Excitation, Light-Quenching.

Current Interest: Spectroscopy, fluorescence and ultrafast time-resolved fluorescence, fluorescence based biomedical sensing, spectroscopy in oriented systems, protein fluorescence and phosphorescence. In particular: FRET, Multi-Photon Excitation, Light Stimulated Emission – Light Quenching, Multi-Pulse Fluorescence.

Four-Photon Excitation of 2,2'-Dimethyl-p-terphenyl, I. Gryczynksi, G. Piszczek, Z. Gryczynski, and J. R. Lakowicz (2002). *J. Phys. Chem. A.,* 106:754-759.

Multiphoton Excitation of Fluorescence near Metallic Particles: Enhanced and Localized Excitation, I. Gryczynski, J. Malicka, Y. Shen, Z. Gryczynski, and J. R. Lakowicz (2002). *J. Phys. Chem. B.* 106:2191-2195.

Gryczynski, Z.
Grygon, C.A.

Date submitted: 29th July 2002

Zygmunt Gryczynski, Ph.D.

Center for Fluorescence Spectroscopy,
Medical Biotechnology Center,
University of Maryland, 725 West Lombard St,
Baltimore, Maryland, 21201, USA.
Tel: 410 706 8409   Fax: 410 706 8409
Karl@cfs.umbi.umd.edu
cfs.umbi.umd.edu

Specialty Keywords: Spectroscopy, Fluorescence, Linear Dichroism, Polarization, Sensing, Protein Ligand Interaction.

Current Interest: Spectroscopy, fluorescence and ultrafast time-resolved fluorescence; application of spectroscopic methods to study biological systems; application of fluorescence to biomedical sensing. In particular FRET, Multi-Photon Excitation, Multi-Pulse Fluorescence, spectroscopy in oriented systems, protein fluorescence and phosphorescence, thermodynamics of protein ligand interaction, fluorescence application to biohazard/bioterrorism and recently metal enhanced fluorescence.

Four-Photon Excitation of 2,2'-Dimethyl-p-terphenyl, I. Gryczynksi, G. Piszczek, Z. Gryczynski, and J. R. Lakowicz (2002). *J. Phys. Chem. A.,* 106:754-759.

Multiphoton Excitation of Fluorescence near Metallic Particles: Enhanced and Localized Excitation, I.

Date submitted: 30th August 2002

Christine A. Grygon, Ph.D.

Boehringer Ingelheim Pharmaceuticals, Inc.,
900 Ridgebury Rd, PO Box 368,
Ridgefield, CT 06877,
USA.
Tel: (203) 798 5651   Fax: (203) 791 6881
cgrygon@rdg.boehringer-ingelheim.com
www.us.boehringer-ingelheim.com

Specialty Keywords: Biophysics, Ligand binding and kinetics, Fluorescence, Anisotropy, Imaging.

Research incorporates the use of multiple biophysics technologies in the validation of molecular and cellular mechanism of action of potential pharmaceutical agents. Fluorescence methods have been employed in the development of assays for high throughput screening, to validate mechanism of action for screening hits, to understand trends in structure-activity relationships, and to study interactions between biological macromolecules.

R.C. Cousins-Wasti, R.H. Ingraham, M.M. Morelock, and **C.A. Grygon**, (1996). Determination of Affinities for *lck* SH2 Binding Peptides Using a Sensitive Fluorescence Assay, *Biochemistry*, **35**, 16746-16752.

Date submitted: 12th August 2002

Oleksiy V. Grygorovych, Ph.D.

Department of Physical Organic Chemistry,
Institute for Chemistry,
Kharkov V. N. Karazin National University,
4 Svobody sqr., Kharkov 61077, Ukraine.
Tel: +38 057 245 7335   Fax: +38 057 245 7130
Alexey.V.Grigorovich@univer.kharkov.ua
www-chemistry.univer.kharkov.ua/dx/nii

Specialty Keywords: Complex formation of organic luminophores, Fluorescent probes.

Current Research Interests: Absorption and fluorescence spectroscopy of conjugated aromatic and heterocyclic organic compounds. Protolytic interactions and complexation with metal ions of conjugated aromatic and heterocyclic organic compounds in their ground and excited states. Photochemical activity of unsaturated organic compounds. Design and application of organic luminophores as new fluorescent probes and sensors for biological systems.

Doroshenko A. O., Grigorovich A. V., Posokhov E. A., Pivovarenko V. G., Demchenko A. P., Sheiko A. D., 2001, Russ. Chem. Bul., **50**, 404-412.

Doroshenko A. O., Sichevska L. V., Grygorovych O. V., Pivovarenko V. G., 2002, Journal of Flourescence, accepted for publication.

Date submitted: 31st August 2002

Yuriy A. Gryzunov, Ph.D.

Russian State Medical University,
Malaya Pirogovskaya 1-A,
Moscow, 119828,
Russia.
Tel: + 7 095 246 4352   Fax: +7 095 246 4512
gryzunov@hotbox.ru

Specialty Keywords: Proteins, Probes, Molecular pathology.

Steady-state and time-resolved spectroscopy, new molecular probes are used to study proteins and lipid-protein complexes both under physiological as well as pathological conditions. Fluorescent probes make it possible to study early changes of conformation and physical-chemical properties of biomacromolecules which are very sensitive to the state of human body.

G.E.Dobretsov, T.I.Syrejshchikova, Yu.A.Gryzunov and M.N.Yakimenko (1998) Quantification of fluorescent molecules in heterogeneous media by use of the fluorescence decay amplitude analysis *J.Fluorescence*, **1**(1), 27-34.

Yu.A.Gryzunov, T.I.Syrejshchikova et al. (2000) Serum albumin binding sites properties in donors and in schizophrenia patients *Nucl. Instr.& Meth.Phys. Res.* **A**(448), 478-482.

Hallberg, E.L.P.
Hamers-Schneider, M.

Date submitted: 9th September 2002

Einar L. P. Hallberg, Ph.D.

Natural Sciences, Södertörns Högskola,
Alfred Nobel's Allé 7, Huddinge,
141 89,
Sweden.
Tel: +46 (8)6084733   Fax: +46 (8)6084510
Einar.hallberg@sh.se
www.sh.se/natur/einar.htm

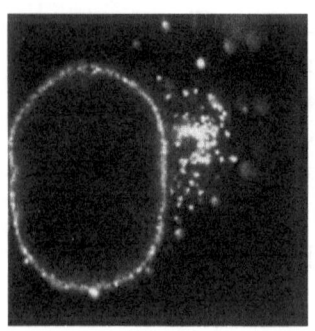

Specialty Keywords: Nuclear membrane, Nuclear pores, GFP.

Trafficking of proteins and RNA molecules in and out of the cell nucleus takes place via the nuclear pore complexes situated in the thousands of pores covering the nuclear surface. We investigate structural and functional aspects of how the nuclear pore complex and the nuclear membranes are organized. We use fluorescence microscopy and confocal laser scanning microscopy on cells expressing proteins tagged with GFP (Green Fluorescent Protein). We also perform Live Cell Imaging including studies of intracellular dynamics using photobleaching.

Kihlmark, M., Imreh, G. and Hallberg, E. (2001) J. Cell Sci., 114, 3643-3653.

Imreh, G. and Hallberg, E. (2000) Exptl. Cell Res., 259, 180-190.

Date submitted: 4th September 2002

Monika Hamers-Schneider, Ph.D.

Chemistry Department, University of Siegen,
D-57068 Siegen, Germany.
&
ATTO-TEC GmbH, D-57008 Siegen, Germany.
Tel: +49 (0) 271 740 4222
monika.hamers@gmx.de, hamers@atto-tec.com
www.atto-tec.com

Specialty Keywords: Fluorescent Dyes, Fluorescent Labels, Fluorescent Sensors.

My research interest is focused on the synthesis of fluorescent labels for bioanalytical applications. Furthermore I am interested in fluorescent dyes which are specially functionalized to meet the requirements of optical sensors.

J. Arden-Jacob, J. Frantzeskos, N.U. Kemnitzer, A. Zilles, and K.H. Drexhage (2001). New fluorescent markers for the red region, *Spectrochim. Acta A*, **57**(11), 2271-2283.
M. Hamers-Schneider (1997). Ph.D.Thesis. Funktionelle Rhodamin-Derivate zur Fluoreszenz-Detektion in Analytik und Sensorik. Shaker Verlag, Aachen.

Date submitted: 22nd August 2002 **Steffen Härtel, Ph.D.**

Departamento de Química Biológica,
Facultad de Ciencias Químicas,
Universidad Nacional de Córdoba,
5000 Córdoba, Argentina.
Tel: 54 351 433 4171 Fax: 54 351 433 4074
shaertel@physik.uni-bremen.de
www.fcq.unc.edu.ar/ciquibic

Specialty Keywords: Lipid Membrane Organization, Image Processing, Fluorescence Microscopy.

Current Research Interests: My current research interest is focused on the development of image processing routines which improve the interpretation of structural and dynamical information, originated in diverse lipid membrane systems. Techniques include fluorescence microscopy of membrane sensitive fluorescent dyes in lipid monolayers, liposomes, and in plasma membranes of living cell cultures.

Härtel, S., Tikhonova, S., Haas, M., Diehl, H. (2002) Membrane Sensitive Fluorescent Dyes for Applications in Fluorescence Microscopy. In press, Journal of Fluorescence.
Fanani, M., Härtel, S., Oliveira, R., Maggio, B. (2002) Bi-directional control of sphingo-myelinase activity and surface topography in lipid monolayers. In press, Biophysical Journal.

Date submitted: 30th May 2002 **Richard P. Haugland, Ph.D.**

Corporate / Research & Development, Molecular Probes Inc.,
4849 Pitchford Avenue, Eugene,
Oregon, 97402,
USA.
Tel: 541 465 8300 Fax: 541 242 0421
richard.haugland@probes.com
www.probes.com

Specialty Keywords: Fluorescence.

Earned Ph.D. in Organic Chemistry from Stanford University (1970). Founder and President of Molecular Probes, Inc. (1975). Author and publisher of the *Handbook of Fluorescent Probes and Research Products*, 9th edition scheduled for release in September 2002. Recent winner (awarded jointly to Dr. Haugland and Dr. Lubert Stryer) of the Molecular Bioanalytics Award 2002 for outstanding achievements in the field of fluorescence resonance energy transfer (FRET).

Date submitted: 30th August 2002

Mary E. Hawkins, M.Sc.

Pediatric Oncology Branch,
National Cancer Institute, National Institutes of Health,
10 Center Drive, Bethesda,
Maryland 20892, USA.
Tel: 301 496 1756 Fax: 301 480 1586
mh100x@nih.gov

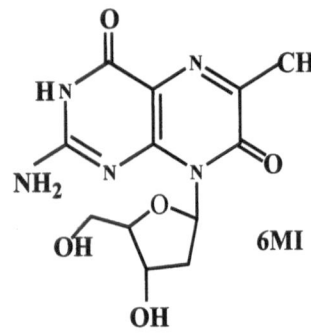

6MI

Specialty Keywords: Pteridine, Nucleoside analog.

We have developed highly fluorescent pteridine nucleoside analog probes synthesized as deoxyribose phosphoramidites ready for site-specific incorporation into oligonucleotides using automated DNA synthesis (TriLink Biotechnologies, San Diego, CA). Native-like linkage positions probes in base-stacked orientation making fluorescence properties exquisitely sensitive to structural changes occurring nearby. Quantum yields for G analogs 3MI & 6MI are 0.9 & 0.7: for A analog, 6MAP, 0.4. These probes are very useful for examination of protein/DNA interactions.

M. E. Hawkins (2001) Fluorescent Pteridine Nucleoside Analogs: A Window on DNA Interactions, *Cell Biochemistry and Biophysics*, **34**, 257-281.

M. E. Hawkins *et al.* (2001) Synthesis and Fluorescence Characterization of Pteridine Adenosine Analogs, *Anal. Biochem.* **298**, 231-240.

Date submitted: 27th August 2002

Michael D. Heagy, Ph.D.

Department of Chemistry, New Mexico Tech,
801 Leroy Ave, Socorro,
Socorro, 87801,
USA.
Tel: 505 835 5417 Fax 505 835 5364
mheagy@nmt.edu
www.nmt.edu/%7Echem/heagy/homepage.html

Specialty Keywords: Chemosensors, Molecular recognition.

The general aims of our ongoing investigations include the application of supramolecular chemistry to the design of fluorogenic reagents for detection of clinically important saccharides. Means of distinguishing between isomers of such biorelevant molecules in protic media (preferably water) at biological pH. Short and practical synthetic methods to fluorescent assemblies that selectively monitor carbohydrates and carbohydrate derivatives. Use of fluorescence spectroscopy to identify signaling mechanisms where substrate binding and fluorescence pathways intersect.

DiCesare, N.; Adhikari, D.P.; Heynekamp, J.J.; Heagy, M.D.; Lakowicz, J.R. *J. Fluor.* **2002**, *12*, 147-154.

Cao, H.; Diaz, D.I; DiCesare, N.; Lakowicz, J.R.; Heagy, M.D. *Org. Lett. 4*, 1503-1505.

Date submitted: 7th September 2002

Ahmed A. Heikal, Ph.D.

School of Applied & Engineering Physics,
Cornell University, 212 Clark Hall,
Ithaca, NY 14583,
USA.
Tel: (607) 255 3919   Fax: (607) 255 7658
aah14@cornell.edu
www.aep.cornell.edu/drbio/drbio.html

Specialty Keywords: Multiphoton fluorescence, Intrinsically fluorescent proteins, Nanoparticles, Molecular dynamics.

I am interested in a molecular-level understanding of complex biological processes. Multidisciplinary approach and integrated fluorescence techniques are critical for such investigation. For example, we have studied the thermodynamics and excited-state fluorescence of green fluorescent proteins (GFP) and red fluorescent protein (DsRed) in both aqueous solution and living cells (1, 2). Furthermore, we have utilized endogenous fluorophores (*e.g.,* NADH and FAD) to monitor the respiratory state activities of cardiac cells using (3) and nervous systems.

A.A. Heikal; S.T. Hess; W.W. Webb; *Chem. Phys.* (2001), 274(1), 37-55.

A.A. Heikal; S.T. Hess; G.S. Baird; R.Y. Tsien; W.W. Webb; *Proc. Natl. Acad. Sci. U. S. A.* (2000), 97(22), 11996-12001.

S. Huang; A.A. Heikal; W.W. Webb; *Biophys. J.* (2002), 82(5), 2811-2825.

Date submitted: 16th September 2002

Sherry L. Hemmingsen, Ph.D.

Varian, Inc.,
2700 Mitchell Drive,
Walnut Creek, CA 94598,
USA.
Tel: (614) 7611330   Fax: (614) 336 0295
sherry.hemmingsen@varianinc.com
www.varianinc.com

Specialty Keywords: Total luminescence spectroscopy, Fluorescence lifetime analysis, Instrumentation.

As Varian's Fluorescence Product Manager for North America, my current efforts focus on supporting a diverse range of customer applications/needs the life sciences, pharma, photonics, etc., plus, contributing to the development of new instrumentation and software.

Former research included fluorescence spectral and lifetime characterization of complex systems such as humic substances, chemometric methods of data analysis, Globals, MEM and total lifetime distribution analysis;

S. L. Hemmingsen and L. B. McGown (1997). Phase-Resolved Fluorescence Spectral and Lifetime Characterization of Commercial Humic Substances: *Appl. Spectrosc.*, **57**, 921.

L. B. McGown, S. L. Hemmingsen, J. M. Shaver, L. Geng (1995). Total Lifetime Distribution Analysis for Fluorescence Fingerprinting and Characterization: *Appl. Spectrosc.*, **49**, 60.

Date submitted: 21st August 2002

Manfred H. Hennecke, Ph.D.

Federal Institute for Materials Research and Testing,
Unter den Eichen 87, Berlin,
D - 12200,
Germany.
Tel: +49 30 8104 1000   Fax: +49 30 8104 1007
hennecke@bam.de
www.bam.de

Specialty Keywords: Fluorescence polarization,
Chemiluminescence.

Physical chemistry of polymers, in particular optical spectroscopy of dimers, oligomers and polymers (especially with polarized light, including time-resolved spectroscopy). Photochemical reactions and aging of polymers (by means of chemiluminescence).

B. Schartel, M. Hennecke, "Thermo-oxidative stability of a conjugated polymer by chemilumi-nescence", Polym. Degr. Stab. **67**, 249-253, 2000.
B. Schartel, S. Krüger, V. Wachtendorf, M. Hennecke, "Excitation energy transfer of a bichromophoric cross-shaped molecule investigated by polarized fluorescence spectroscopy"
J. Chem. Phys. **112**, 9822-9827, 2000.

Date submitted: 9th September 2002

Albin Hermetter, Ph.D.

Department of Biochemistry,
Technische Universität Graz,
Petersgasse 12/2, A-8010 Graz,
Austria.
Tel: +43 316 873 6457   Fax: +43 316 873 6952
albin.hermetter@tugraz.at
www.biochemistry.tugraz.at/

Specialty Keywords: Lipoproteins, Lipases, Membranes.

Our research deals with the role of glycero(phospho)lipids as components of membranes and lipoproteins, as mediators in cellular (patho)biochemistry, and their application as substrates and analytical tools in enzyme technology. We develop and apply fluorescence techniques to study lipid oxidation, the effects of oxidized lipids on intracellular signalling, the inhibition of these processes by antioxidants, and function of lipolytic enzymes in biocatalysis and medicine.
Fluorescent inhibitors for the qualitative and quantitative analysis of lipolytic enzymes.
H. Scholze, H. Stütz, F. Paltauf, and A. Hermetter, Anal. Biochem 276, 72 - 80 (1999).
High-throughput fluorescence screening of antioxidative capacity in human serum.
Mayer, M. Schumacher, H. Brandstätter, F.S. Wagner, and A. Hermetter, Anal.Biochem. 297, 144 – 153 (2001).

Date submitted: 10th September 2002 **Andreas Herrmann, Ph.D.**

Institute of Biology, Humboldt-University Berlin,
Invalidenstr. 42, Berlin,
D-10115,
Germany.
Tel: 49 30 2093 8860   Fax: 49 30 2093 8585
andreas.herrmann@rz.hu-berlin.de
www.biologie.hu-berlin.de/~molbp/new/

Specialty Keywords: Membrane, Fusion, Flip-flop.

The research focuses on the following topics: (i) transport of lipids across biological membranes, (ii) the trafficking of lipids in eukaryotic cells, (iii) protein-mediated fusion of biological membranes, and (iv) protein-lipid interaction. Various spectroscopical methods including fluorescence spectrosocopy and quantitative fluorescence microscopy are employed. (Fluorescent) labeling of biological molecules is achieved by molecular biology approaches (proteins) or by chemical synthesis (lipid analogues).

Kubelt, J., AK. Menon, P. Müller, A. Herrmann (2002) *Biochemistry*. **41**, 5605-5612.

John, K., J. Kubelt, P. Müller, D.Wüstner, A. Herrmann (2002) *Biophys. J.* **83**, 1525-1534.

Date submitted: 11th September 2002 **Joseph D. Hewitt, Ph.D.**

Varian Analytical Instruments,
2700 Mitchell Dr., Walnut Creek,
CA, 94598,
USA.
Tel: 1 800 926 3000 ext 3064   Fax: 925 945 2360
Joe.Hewitt@varianinc.com
www.varianinc.com

Specialty Keywords: Fluorescence Instrumentation.

Current Research Interests: As a fluorescence product specialist for Varian in the Midwest US, I work on application questions, Eclipse spectrofluorometer demonstrations and sales support. My individual research interests include humic substance lifetime spectroscopy, coupled detection schemes and fluorescence sensing technology.

Date submitted: 11th September 2002

Andrew R. Hind, Ph.D.

UV-Vis-NIR Sales Support Manager Europe,
Varian, Ltd., 28 Manor Road, Walton-on-Thames,
Surrey, KT 12 2Qf,
England.
Tel: + 44 (0) 1932 898000   Fax: +44 (0) 1932 228769
andrew.hind@varianinc.com
www.varianinc.com

Specialty Keywords: Materials science, Industrial chemistry, Optics / photonics, Molecular spectroscopy.

Background in 'applied' molecular spectroscopy research, with focus on applications in the materials science, industrial chemistry and optics/photonics areas. Experienced in the use of fluorescence, UV-Vis (including far-UV), infrared (near-, mid-, and far-) and Raman spectroscopies, with particular areas of interest including semiconductor, telecommunications, mineralogical and coating/surface characterization applications.   Very interested in new spectroscopic instrumentation, techniques and applications.

A.R. Hind, S.K. Bhargava, and S.C. Grocott (1999) Colloids Surf. A. 146, 359-374.
A.R. Hind, S.K. Bhargava, and A. McKinnon (2001) Adv. Colloid Interfac. Sci. 93, 91-114.

Date submitted: 17th June 2002

Rhoda Elison Hirsch, Ph.D.

Dept of Medicine (Hematology) and Dept of Anatomy & Structural Biology, Albert Einstein College of Medicine, 1300 Morris Park Avenue, Bronx, NY 10579, USA
Tel: 718 430 3604   Fax: 718 824 3153
rhirsch@aecom.yu.edu

Specialty Keywords: Hemoglobin, Front-face fluorometry, Hemoglobin C crystal growth.

Our laboratory focuses on the ß6 hemoglobin mutants that form aggregates in the red blood cell: Why does oxy HbC (ß6 Glu ✹ Lys) form crystals in the red blood cell in contrast to deoxy sickle cell hemoglobin [HbS, ß6 Glu ✹ Val] that forms polymers? We are also interested in model hemoglobin based blood substitutes and stabilization mechanisms. The application of front-face fluorescence to study hemoglobin and heme proteins is ongoing in the laboratory.

RE Hirsch, "Heme Protein Fluorescence". Chapter 10 (pp. 221-255) Topics in Fluorescence Spectroscopy, Volume 6, Protein Fluorescence (ed. JR Lakowicz), New York (2000)
QY Chen, C Bonaventura, RL Nagel, and RE Hirsch. "Distinct Domain Responses of R-state Human Hemoglobins A, C, and S to Anions." Blood Cells, Molecules and Diseases (2002), in press.

Date submitted: 12th August 2002   **Martin Hof, Ph.D.**

Center for Complex Molecular Systems and Biomolecules,
J. Heyrovský Institute of Physical Chemistry,
Academy of Sciences of the Czech Republic,
Dolejškova 3, Cz-18223 Prague 8, Czech Republic.
Tel: +420 266 053 264   Fax: +420 286 582 307
Hof@jh-inst.cas.cz
www-troja.fjfi.cvut.cz/k412/home/hof/en/index.html

Specialty Keywords: Solvent Relaxation, Tryptophan Fluorescence, Fluorescence Correlation Spectroscopy (FCS).

Following main topics are presently pursued in M. Hof's laboratory:

1) Solvent relaxation in phospholipid bilayers [1]: Basic principles, applications, and new membrane labels.
2) FCS as a tool for the characterization of DNA condensation [2].
3) Formation of phospholipid mono- and bilayers controlled by FCS.
4) Picosecond tryptophan fluorescence of blood coagulation proteins.

[1] J. Sýkora, P. Kapusta, V. Fidler, M. Hof On What Time-Scale Does Solvent Relaxation in Phospholipid Bilayers Happen? (2002), Langmuir, 18(3), 571-574.

[2] T. Kral, M. Hof, M. Langner Effect of Spermine on the Plasmide Condensation and Dye Release Observed by FCS (2002), Biol. Chem. 383 (2), 331-335.

Date submitted: 13th September 2002   **Johannes W. Hofstraat, Ph.D.**

Dept. of Polymers & Organic Chemistry, Philips Research,
Prof. Holstlaan 4, 5656 AA Eindhoven,
Institute of Molecular Chemistry,
University of Amsterdam,
The Netherlands.
Tel: +31 40 2744910   Fax: +31 40 2743350
hans.hofstraat@philips.com

Specialty Keywords: Materials, Displays, Diagnostics, Photonics.

Research topics: (Electro)luminescent polymers, dyes, in particular luminescent metal complexes, and self-organizing materials, for application in displays (emissive, liquid crystalline, reflective), storage (optical, solid-state), electronics (mainly polymer-based) and sensors, e.g. for medical applications (diagnostics, imaging). Research on (opto-)electronic devices: preparation and characterization. Advanced instrumentation for ultra fast time-resolved measurements, for microscopy and for imaging, also for near-infrared luminescence.

K. Brunner, J.A.E.H. van Haare, B.M.W. Langeveld-Voss, H.F.M. Schoo, J.W. Hofstraat, A. van Dijken, J. Phys. Chem. B, 106, 6834-6841 (2002).

L.H. Slooff, A. van Blaaderen, A. Polman, G.A. Hebbink, S.I. Klink, F.C.J.M. van Veggel, D.N. Reinhoudt, J.W. Hofstraat, J. Appl. Phys., 91, 3955-3980 (2002)

Date submitted: 8th August 2002

Elisabeth Klara, Holder. Ph.D.

Center for Fluorescence Spectroscopy,
Medical Biotechnology Center,
University of Maryland, 725 West Lombard St,
Baltimore, Maryland, 21201, USA.
Tel: 410 706 7500   Fax: 410 706 8408
elisabeth@cfs.umbi.umd.edu
cfs.umbi.umd.edu

Specialty Keywords: Macromolecular Dynamics, Fluorescence Sensing, and Principles of RDE.

Fluorescence methods offer a powerful tool to study dynamics of polymers and biopolymers and in fluorescence sensing. These approaches involve the design of new probes with improved characteristics. RDE with all exciting effects requires a detailed theoretical study of metal surface modified fluorescence.

H.–J. Egelhaaf, E. Holder, P. Herman, H. A. Mayer, D. Oelkrug, and E. Lindner (2001). Synthesis, characterisation, and fluorescence spectroscopic mobility studies of fluorene labeled inorganic–organic hybrid polymers *Journal of Materials Chemistry* **11**, 2445-2452.

E. Holder, J. R. Lakowicz, N. DiCesare, K. Briggman, I. Gryczynski, J. Malicka, Y. Shen, and Z. Gryczynski *Intensified fluorescence by metal modified radiative rates*
224th ACS National Meeting, Boston, Massachusetts, USA. August 18–22, 2002.

Date submitted: 18th August 2002

Bonnie J. Howell, Ph.D.

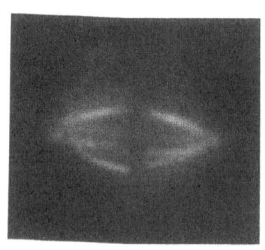

Biology, University of North Carolina,
CB#3280, 607 Fordham Hall, Chapel Hill,
Orange, 27599,
USA.
Tel: 919 962 2354   Fax: 919 962 1625
bhowell@email.unc.edu
www.unc.edu/%7Ebhowell/

Specialty Keywords: Fluorescence, FRAP, Spindle checkpoint.

The spindle checkpoint prevents aneuploidy by inhibiting anaphase onset until all chromosomes have achieved proper spindle attachment and alignment. To elucidate a mechanism for spindle checkpoint activity, I've used quantitative fluorescence, phase-contrast, and confocal microscopy to examine the localization pattern and dynamic behavior of spindle checkpoint proteins in living mammalian tissue culture cells. Fluorescence recovery after photobleaching (FRAP) techniques have also been used to determine the transitory nature of these components at kinetochores and spindle poles.

Howell, B.J. et al. 2000. J. Cell Biol. *150*: 1233-1249.
Howell, B.J. et al. 2001. J. Cell Biol. *155*: 1159-1172.

Date submitted: 7th July 2002

Graham Hungerford, Ph.D.

Departamento de Física,
Universidade do Minho,
4710-057 Braga,
Portugal.

graham@fisica.uminho.pt

Specialty Keywords: Time-resolved fluorescence, Sol-gel and Microheterogeneous systems.

My present research interests involve the manufacture and study (using fluorescence techniques) of sol-gel-derived matrices to elucidate dye-dye and dye-host interactions. The matrices are made using either Si or Ti precursors to form "passive" or "active" hosts, in which we can incorporate solvatocromic probes, porphyrins and phthalocyanines. Similar probes have also been employed in a study using surfactant systems.

G. Hungerford, J.A. Ferreira (2001). The effect of the nature of retained solvent on the fluorescence of nile red incorporated in sol-gel-derived matrices. *J. Lumin.* **91**, 155-165.

G. Hungerford *et al.* (2002). Monitoring ternary systems of $C_{12}E_5$/water/tetradecane via the fluorescence of solvatochromic probes. *J. Phys. Chem. B.* **106**, 4061-4069.

Date submitted: 6th September 2002

Takamitsu Ikkai, Ph.D.

Biophysics, Aichi Prefectural University of Fine Arts,
Sagamine 1-1, Nagakute,
Aichi, 480-1194,
Japan.
Tel: +81 561 62 1180 Fax: +81 593 31 3406
ikkai@mail.aichi-fam-u.ac.jp
www.aichi-fam-u.ac.jp/~ikkai

Specialty Keywords: Fluorometric search of protein conformation, Bioluminescence, Crystallization.

Using fluorescence, we are studying a phase transition of proteins. Protein crystallization is one of a phase transition induced by precipitants. Fluorescence monitoring based on a short column method has been successful for the search of nuclei to get crystals of actin filaments. This method is effective to find crystallization buffers quickly. Applications of this technique for crystallization of other proteins and their complexes are now in progress.

T. Ikkai and K. Shimada, A Flow System Based on a Fluorometer and a Luminometer to Monitor the Correlation of Protein Conformation and Function. J. Fluoresc 10, 77-79 (2000).
T. Ikkai and K. Shimada, Introduction of Fluorometry to the Screening of Protein Crystallization Buffers. J. Fluoresc 12, 167-171 (2002).

Ito, A.S.
Jankowski, A.

Date submitted: 30th August 2002 **Amando S. Ito, Ph.D.**

Departamento de Física e Matemática,
FFCLRP – Universidade de São Paulo,
Av. Bandeirantes 3900, Ribeirao Preto,
14015 120, Brazil.
Tel: 55 16 302 3693
amando@dfm.ffclrp.usp.br
dfm.ffclrp.usp.br/fotobiofisica/membros.html

Specialty Keywords: FRET, Peptide conformational dynamics, Peptide / lipid interaction.

Research interests: physico-chemical properties of extrinsic and intrinsic fluorescent probes for peptides and proteins. Donor-acceptor distance distribution and conformational dynamics in peptides. Labeled macromolecules in interaction with supramolecular assemblies. Fluorescence studies on membrane models.

A.S. Ito, E.S. Souza, S.R. Barbosa and C.R. Nakaie. (2001) Fluorescence Study of Melanotropins Labeled with Aminobenzoic Acid. *Biophysical Journal*, 81, 1180-1189.
D.C.Pimenta, I.L.Nantes, E.S.Souza, B. le Boniec, A.S.Ito, I.L.S.Tersariol, V.Oliveira, M.A.Juliano and L.Juliano. (2002) Interaction of heparin with internally quenched fluorogeic peptides. *Biochem. J.*, 366, 435-446.

Date submitted: 24th August 2002 **Andrzej Jankowski, Ph.D.**

University of Zielona Gora.Institute of Biotechnology and Environment Protection,
Monte-Cassino 21b Zielona Gora,
Poland 65-561,
Poland.
Tel: (048) (071) 3539177
JJJ@WCHUWR.CHEM.UNI.WRIC.PL

Specialty Keywords: Fluorescence, Bioorganic Chemistry, Envioronment Chemistry.
The main topics in scientific activity: Kinetics and equilibria of enzyme denaturation. The structure of peptides in solution. Excited-state proton transfer. Phiotosensitization of bacteria by natural and synthetic dyes. The most outstanding achievements: the description of the fluorescence of the anilinium cation of p-amino phenylalanine. explaining the enhancement of the proton transfer rate in naphthol derivatives bound to the active site of papain. The determination of intramolecular distances in peptides affeting central nervous system .

A.Jankowski and P.Dobryszycki: Spectroscopic properties of 4-amino phenylalanine.Photochem.Photobiol.44(1986)159-168.
A.Jankowski and P.Stefanowicz: Investigations of excited state proton transfer in 2-naphthol derivativesbound to selected sites of proteins.J.Photochem.Photobiol A.Chem 69(1992)57-66 see also ibidem.84(1994)143-140 and 85(1995)69-75.

Date submitted: 2nd July 2002

Lennart B.-Å Johansson, Ph.D.

Department of Chemistry,
Biophysical Chemistry,
Umeå University,
S-901 87 Umeå, Sweden.
Tel: +46 (0) 90 786 5149   Fax: +46 (0) 90 786 7779
Lennart.Johansson@chem.umu.se
www.umu.chem.se/ljn/

Specialty Keywords: Energy transfer / migration, Polarised emission, Structure of biomacromolecules.

My research mainly focuses on developing new versatile tools for examining biomacromolecular structure-function, mainly of proteins. For this methods and theory are developed based on donor-donor energy migration. Derivatives of BODIPY are extensively used and characterized. Recent work reveals that BODIPY exhibits spectacular properties of dimerisation that enable new applications. Studies based on two-photon excited fluorescence are in progress.

F. Bergström, P. Hägglöf, J. Karolin, T. Ny, and L. B.-Å. Johansson (1999) JACS, **96**, 12477-12481.

F. Bergström, I. Mikhalyov, P. Hägglöf, R. Wortmann., T. Ny, and L. B.-Å. Johansson (2002) PNAS, **124**, 196-202.

Date submitted: 27th August 2002

Arthur E. Johnson, Ph.D.

Wehner-Welch Chair, Dept. Medical Biochemistry & Genetics,
Texas A&M Univesity System Health Science Center,
116 Reynolds Medical Building, 1114 TAMU,
College Station, TX 77843-1114 USA.
Tel: (979) 862 3188 Fax: (979) 862 3339
aejohnson@tamu.edu
tamushsc.tamu.edu/aejohnson/home/index.html

Specialty Keywords: Protein-Membrane Interactions, Fluorescence, FRET.

We are investigating the movement of proteins through or into a membrane (protein sorting), the creation of holes in mammalian cell membranes by bacterial toxins, blood coagulation, and protein biosynthesis. Various fluorescence techniques are used to characterize the molecular interactions and conformational changes involved in the assembly, function, and regulation of membrane-bound protein complexes. FRET is used to determine their structure and topography.

N. G. Haigh, and A. E. Johnson (2002) A New Role for BiP: Closing the Aqueous Translocon Pore during Protein Integration into the ER Membrane, *Journal of Cell Biology* **156**, 261-270.

A. P. Heuck, E. M. Hotze, R. K. Tweten, and A. E. Johnson (2000) Mechanism of Membrane Insertion of a Multimeric Barrel Protein: Perfringolysin O Creates a Pore Using Ordered and Coupled Conformational Changes, *Molecular Cell* **6**, 1233-1242.

Date submitted: 8th September 2002

Carey K. Johnson, Ph.D.

Department of Chemistry, University of Kansas,
1251 Wescoe Hall Drive, Lawrence,
KS, 66045-7582,
USA.
Tel: 785 864 4219   Fax: 785 864 5396
ckjohnson@ku.edu
www.chem.ku.edu/CJohnson/Default.asp

Specialty Keywords: Time-resolved fluorescence, Single-molecule fluorescence.

The research in my laboratory focuses on the dynamic properties peptides and proteins by time-resolved and single-molecule spectroscopy. We use single-molecule fluorescence spectroscopy to investigate calcium signaling by single calmodulin molecules. In another project, time-resolved fluorescence anisotropy and resonance energy transfer experiments are being used to probe the dynamics of short peptides in solution. We also study ultrafast processes in chromophores of biological importance.

C. K. Johnson and C. Z. Wan (1997). Anisotropy decays induced by two-photon excitation. *Topics in Fluorescence Spectroscopy* **5** (Multiphoton Excitation and Light Quenching) J. R. Lakowicz, ed., (Plenum, New York, 1997) 43-85.

Date submitted: 24th May 2002

Michael L. Johnson, Ph.D.

Department of Pharmacology and Internal Medicine,
University of Virginia, Box 800735,
Charlottesville,
VA 22908-0735, USA.
Tel: 434 924 8607   Fax: 434 982 3878
mlj8e@virginia.edu
www.med.virginia.edu/medicine/basic-sci/pharm/johnson.html

Specialty Keywords: Mathematical Modeling, Biophysics.

My research interests center on understanding the biochemical, physical chemical, and thermodynamic pathways by which one portion of a biological organism transfers information to other portions of the same organism. This interest has spawned research at the level of whole organisms, at the cellular level and at the molecular level.

Date submitted: 30th August 2002   **Anita C. Jones. Ph.D.**

School of Chemistry & Collaborative Optical Spectroscopy,
Micromanipulation and Imaging Centre,
University of Edinburgh,
Edinburgh, EH9 3JJ, UK.
Tel: +44 (0) 131 650 6449   Fax: +44 (0) 131 650 4743
a.c.jones@ed.ac.uk
www.chem.ed.ac.uk and www.cosmic.ed.ac.uk

Specialty Keywords: Spectroscopy, Photophysics, Time-resolved fluorescence, FLIM.

Research interests: Steady state and time-resolved fluorescence spectroscopy; fluorescence lifetime imaging; molecular photophysics and photochemistry; use of fluorescence to probe biomolecular systems; photophysics of luminescent polymers; industrial and biomedical applications of fluorescence.

N.M. Speirs , W.J. Ebenezer and A.C. Jones (2002). Observation of a fluorescent dimer of a sulfonated phthalocyanine, *Photochem.Photobiol* **76**, 247-251.
A C Jones, M. Millington, J Muhl, J M De Freitas, J S Barton and G Gregory (2001). Calibration of an optical fluorescence method for film thickness measurement, *Measurement Science & Technology*, **12**, N23-N27.

Date submitted: 22nd August 2002   **Hans-Peter Josel, Ph.D.**

Roche Diagnostics GmbH,
R&D Diagnostics,
Nonnenwald 2, 823372 Penzberg,
Germany.
Tel: +49 8856 60 5289   Fax: +49 8856 60 5311
hans-peter.josel@roche.com
www.roche.com

Specialty Keywords: Fluorescence Label, Time Resolved Fluorescence Label, FRET.

Development of tailor made luminescence label, esp. fluorescence label, time resolved fluorescence label, specially designed FRET systems for use in detection in diagnostics (DNA and others) and pharma screening.

Josel, Hans-Peter; Herrmann, Rupert; Heindl, Dieter; Muehlegger, Klaus; Sagner, Gregor; Drexhage, Karl Heinz; Frantzeskos, Jorg; Arden-Jacob, Jutta. Fluorescent rhodamine dye derivatives and their use in diagnostic systems.   Eur. Pat. Appl. (1999) EP 962497
Herrmann, Rupert; Josel, Hans Peter; Drexhage, Karl Heinz; Arden, Jutta. Pentacyclic compounds, their use as dyes and fluorescent dyes, and immunoassay therewith.   Ger. Offen. (1993), DE 4137934

Kang, H.C.
Kang, J.S.

Date submitted: 31st May 2002

Hee Chol Kang, Ph.D.

Organic Chemistry, Molecular Probes, Inc.,
4849 Pitchford Avenue, Eugene,
Oregon, 97402,
USA.
Tel: 541 465 8300   Fax: 541 344 6504
heechol.kang@probes.com
www.probes.com

Specialty Keywords: Fluorescence, Time-resolved probes, Nucleotides.

My research focuses on the design and development of novel fluorescent molecules with improved spectral properties. Recent projects include: design and synthesis of novel fluorescent organometallic complexes with long fluorescence lifetimes and high Stokes shift for time-resolved applications, design and synthesis of fluorescent probes for direct chemical labeling of nucleic acids, and, synthesis of a wide variety of fluorescent nucleotides for the study of nucleotide-binding proteins.

Date submitted: 9th July 2002

Jung Sook Kang, Ph.D.

Dept. of Oral Biochemistry and Molecular Biology,
College of Dentistry, Pusan National University,
1-10 Ami-dong, Su-gu, Pusan 602-739,
Korea.
Tel: 82 51 240 7820   Fax: 82 51 241 1226
jsokang@pusan.ac.kr

Specialty Keywords: Macromolecular dynamics, Frequency-domain fluorometry, Long-lifetime metal-ligand complex.

I have been studying the dynamics of proteins, nucleic acids, and membrane lipids using a variety of fluorescence techniques. Recently my research was focused on investigating macromolecular dynamics using long-lifetime metal-ligand complexes.

Kang J. S., Piszczek G. and Lakowicz J. R. (2002) High-molecular-weight protein hydrodynamics studied with a long-lifetime metal-ligand complex. *Biochim. Biophys. Acta* 1597, 221-228.
Kang J. S., Abugo O. O. and Lakowicz J. R. (2002) Dynamics of supercoiled and liner pTZ18U plasmids observed with a long-lifetime metal-ligand complex. *Biopolymers* 67, 121-128.

Date submitted: 6th September 2002

András D. Kaposi, Ph.D.

Dept. of Biophysics and Radiation Biology,
Semmelweis University,
Pusin u. 9, Budapest, H-1088,
Hungary.
Tel: (36 1) 266 2755 / 40 33   Fax: (36 1) 266 6656
kaposi@puskin.sote.hu

Specialty Keywords: Fluorescence line narrowing, Spectroscopy inhomogeneous broadening.

Research interests: laser excited high-resolution luminescence spectroscopy of inhomogeneously broadaned samples, energy selective optical spectroscopy of heme proteins, understanding of factors that influence the fluorescence line narrowing spectra, substrate binding to heme proteins (fluorescence, FTIR and visible absorption spectroscopy), natural chromophores, plant and bacterial fluorescence.

Fidy J., M. Laberge, A.D. Kaposi and J.M. Vanderkooi, (1998). Fluorescence line narrowing applied to the study of proteins *Biochim. Biophys. Acta* **1386**, 331-351.

Kaposi A.D., W.W. Wright and J.M. Vanderkooi, (2002). Consequences of inhomogeneous broadening on fluorescence line narrowing spectra *J. Fluorescence* (accepted).

Date submitted: 24th June 2002

Peter Kapusta, Ph.D.

PicoQuant GmbH,
Rudower Chaussee 24,
Berlin 12489,
Germany.
Tel: +49 (0) 30 63926914   Fax: +49 (0) 30 63926561
kapusta@pq.fta-berlin.de
www.picoquant.com

Specialty Keywords: Time-resolved Spectrometer Design, TCSPC, Laser diode, LED, Anisotropy, Solvation.

Current Status and Scientific Interests: Senior scientist at PicoQuant GmbH, development of laser diode and LED based time-resolved fluorescence instrumentation, promotion of the application of the TCSPC method in various research areas, photophysics of novel fluorophores, energy transfer in bichromophores, solvation dynamics, computational chemistry.

Date submitted: 3rd September 2002

Peet Kask, Ph.D.

Applied Biophysics, Evotec OAI AG,
Instituudi tee 11, Harku,
Harku 76902, Harjumaa,
Estonia.
Tel / Fax: +372 656 0601
peet@ebi.ee

Specialty Keywords: FCS, FIDA, Photon Statistics.

Development of new fluorescence methods of a single molecule sensitivity for applications in e.g. high throughput drug screening. Molecular species are recognized on basis of fluorescence lifetime, specific brightness, fluorescence anisotropy, diffusion time and other specific molecular properties. The so-called FIDA-family of histogram methods has been developed: FIDA, 2D-FIDA, FIMDA and FILDA.

K.Palo, L.Brand, C.Eggeling, S.Jäger, P.Kask and K.Gall. Fluorescence intensity and lifetime distribution analysis: Toward higher accuracy *Biophys.J.* (2002) 83(2), 605-618.
P.Kask, K.Palo, N.Fay, L.Brand, Ü.Mets, D.Ullmann, J.Jungmann, J.Pschorr, and K.Gall. Two-dimensional fluorescence intensity distribution analysis *Biophys.J.* (2000) 78(4), 1703-1713.

Date submitted: 4th September 2002

Norbert U. Kemnitzer, Ph.D.

ATTO-TEC GmbH,
D-57008 Siegen,
Germany.
Tel: +49 (0) 271 740 4222

norbert.kemnitzer@gmx.de, kemnitzer@atto-tec.com
www.atto-tec.com

Specialty Keywords: Organic Dyes, Fluorescent Labels.

My research interest is the development and synthesis of fluorescent dyes suitable as labels for applications in biochemistry and medicine. Therefore I am particularly interested in the design and chemical modification of chromophoric systems for the red region of the visible spectrum.

J. Arden-Jacob, J. Frantzeskos, N.U. Kemnitzer, A. Zilles, and K.H. Drexhage (2001). New fluorescent markers for the red region, *Spectrochim. Acta A,* **57**(11), 2271-2283.
N.U. Kemnitzer (2001). Ph.D.Thesis. Amidopyrylium-Fluoreszenz-Farbstoffe. Der Andere Verlag, Osnabrück.

Date submitted: 30th August 2002

Borys Kierdaszuk, Ph.D., D.Sc.

Laboratory for Fluorescence Spectroscopy of Biological,
Molecules, Dept. of Biophysics, Inst. of Exptl Physics,
University of Warsaw, Zwirki i Wigury 93,
PL-02089 Warsaw, Poland.
Tel: +(004822) 5540715 / 5540727   Fax: +(004822) 5540001
borys@biogeo.uw.edu.pl
www.fuw.edu.pl

Specialty Keywords: Emission spectroscopy of biological molecules, Protein-ligand interactions, Fluorescence probes.

Emission (fluorescence, phosphorescence) spectroscopy applicable to biophysical studies of bio-macromolecular systems and their constituents, e.g. mechanism of recognition and kinetics of protein-ligand binding, identification and characterization of reaction transition states, confrontation of crystallographic data with solution studies; to better understand the mechanisms of catalysis, towards development of sensitive and selective methods of detection.

Kierdaszuk B., Modrak-Wójcik A., Wierzchowski J., Shugar D. (2000) Induced tautomeric shifts on binding to enzyme, and enzyme-ligand FRET. *Biochim. Biophys. Acta* **1476**, 109-128.
Stepanenko T., Lapinski L., Sobolewski A.L., Nowak M.J., Kierdaszuk B. (2000) Photochemical syn-anti isomerisation reaction. *J. Phys. Chem.* **104**, 9459-9466.

Date submitted: 30th August 2002

Paavo K.J. Kinnunen, Ph.D.

Memphys – Center for Biomembrane Physics,
Helsinki Biophysics & Biomembrane Group,
Department of Medical Chemistry, Institute of Biomedicine,
POB 63, FIN-00014 University of Helsinki, Finland.
Tel: +358 9 19125400   Fax: +358 9 19125444
Paavo.Kinnunen@Helsinki. Fi

Specialty Keywords: Lipids, Biomembranes, Lipid-protein interactions.

The major line of research of HBBG pursues the molecular level mechanisms underlying both 2-D and 3-D ordering of supramolecular assemblies constituted by lipids, aiming to compile an integrated view on the coupling between the physical properties of lipids to the physiological functions of biomembranes. More specifically, we are elucidating the mechanisms which convey changes in the physicochemical characteristics of bilayer lipids to the conformation and activity of membrane proteins.

Kleszczyńska, H.
König, K.

Date submitted: 12th September 2002

Halina Kleszczyńska, Ph.D.

Department of Physics and Biophysics,
Agricultural University,
Wrocław, Norwida 25, 50-375,
Poland.
Tel: +48 713205141   Fax: +48 713205172
halina@ozi.ar.wroc.pl
www.ar.wroc.pl

Specialty Keywords: Erythrocyte, Membrane, Fluorescence anisotropy.

In the Department of Physics and Biophysics of Agricultural University in Wrocław Poland I conduct investigations on the molecular mechanism of interaction between membrane-active and antioxidative substances and erythrocyte membranes. We apply the fluorimetric method to study the fluidity of erythrocyte membranes, based on the coefficients of fluorescence polarization and/or anisotropy, using the fluorescent probes (DPH, TMA-DPH and TMAP-DPH), The measurements were performed with SFM 25 spectrofluorometer (KONTRON, Zurich, Switzerland).

Date submitted: 28th August 2002

Karsten König, Ph.D.

Center for Lasermicroscopy, University Jena,
Teichgraben 7,
Jena, 07743,
Germany.
Tel: +49 3641 938560   Fax: +49 3641 938590
kkoe@mti-n.uni-jena.de
www.uni-jena.de/clm

Specialty Keywords: Multiphoton microscopy, Time-resolved single photon counting, Tissue imaging, Autofluorescence.

Research is focussed on multiphoton fluorescence microscopy and imaging of tissue autofluorescence with high submicron spatial resolution, 250 ps temporal resolution and 5 nm spectral resolution. Our studies include the single/few molecule level (e.g. Multiphoton Multicolor FISH, time-resolved FRET), the single cell level (e.g. GFP expression after optical gene transfer, imaging of optically trapped gametes and microorganisms) and *in vivo* studies on tissues (optical multiphoton tomography of skin and eyes). The equipment includes femtosecond laser scanning microsopes, a TauMap microscope for fluorescence lifetime imaging, systems for nanosurgery and imaging, laser tweezers and the multiphoton skin imaging system DermaInspect 100.

König: Review. Multiphoton microscopy in life sciences. J. Microsc. 200(2000)83-104
Tirlapur, König: Targeted transfection by femtosecond laser. Nature. 418(2002)290-291.

Date submitted: 10th September 2002

Vishwanath Koppaka, Ph.D.

Department of Pharmacology,
University of Pennsylvania,
3610 Hamilton Walk, 105 Johnson Pav,
Philadelphia, PA 19104, USA.
Tel: 215 573 7567   Fax: 215 573 2236
koppaka@mail.med.upenn.edu

Specialty Keywords: Fluorescent probes, Lipoprotein A-I, Lipid oxidation.

Research involves investigating protein-lipid interactions and their effect on structure/function using biophysical techniques such as polarized attenuated total internal reflectance Fourier transform infrared (PATIR-FTIR) and fluorescence spectroscopies. Current focus is on investigation of structure, orientation, and interaction of the protein and lipid components in high-density lipoprotein particles and the effect of oxidatively damaged lipids and acute phase response proteins (injury specific apolipoproteins) on reverse cholesterol transport.

Koppaka, V. Structural Studies of Discoidal Lipoprotein A-I. Cellular and Molecular Life Sciences. 58: 885-893, 2001.

Date submitted: 10th September 2002

Valentin I. Korotkov, Ph.D.

General Physics 2, Department. of Physics,
St. Petersburg State University,
Ulianovskaya, I, Petrodvorets, St. Petersburg,
198504, Russia.
Tel: +7 (812) 428 4366 Fax: +7 (812)428 7240
korotkov@paloma.spbu.ru

Specialty Keywords: Adsorption, Sensitization, Energy transfer.

Two quantum processes in photosensitized decomposition of water: 1/ as a result of energy transfer from high triplet levels of organic molecules (naphthalene, biphenyl, benzene) adsorbed on silica to dissociative triplet levels of water; 2/ promoted with the absorbed molecules of phthalocyanine and p-benzoquinone via formation of dark charge transfer complexes[1]. Studying of luminescence of surface molecules of various organic molecular crystals in comparison with the luminescence of bulk molecules of the same crystals [2].

A.V. Barmasov, V.I. Korotkov, V.Y. Kholmogorov (1994). Model photosynthetic system with charge transfer for transforming solar energy. *Biophysics*. **39**(2), 227-231.

E.P. Zarochentseva, V.I. Korotkov, Ya. P. Oleinik, V.Y. Kholmogorov (1996). Luminescence of benzoik acid polycrystals doped with bromated diphenils. *Optics and Spectros*.**81**(4), 570-573.

Date submitted: 29th August 2002

Yurii V. Korovin, Ph.D.

A.V. Bogatsky Physico-Chemical Institute,
of the National Academy of Sciences of Ukraine,
86 Lustdorfskaya doroga, Odessa,
65080, Ukraine.
Tel: +38 (0482) 652 380   Fax: +38 (0482) 652 012
physchem@paco.net

Specialty Keywords: Lanthanides, IR-luminescence, Macrocyclic ligands.

Current Research Interests: Design and investigation of lanthanide complexes with macrocyclic ligands of different types (e.g. porphyrines, calixarenes, ligands bearing aromatic antennae). New types of lanthanide complexes for use in biomedicine, in particular, as IR-luminescent markers.

Yu.Korovin and N.Rusakova (2002). Near Infrared Luminescence of Lanthanides in Complexes with Organic Dyes. *J. Fluorescence*. **12**, 159-161.
Yu.Korovin and N.Rusakova (2001). Infrared 4f-Luminescence of Lanthanides in the Complexes with Macrocyclic Ligands. *Rev. Inorg. Chem.* **21**, 299-329.

Date submitted: 30th August 2002

Vladyslava B. Kovalska, Ph.D.

Institute of Molecular Biology and Genetics of NAS of Ukraine, Zabolotnogo Str. 150,
Kyiv,
03143, Ukraine.
Tel: (+38 044) 252 23 89   Fax: (+38 044) 252 24 58
pelikash@svitonline.com
www.yarmoluk.org.ua

Specialty Keywords: Fluorescent probes, Cyanine dyes, Nucleic acids.

The research activities of Dr. V.Kovalska are aimed on the designing of fluorescent probes for nucleic acid and protein detection [1]. Now she is working as Scientific Researcher in Nucleic Acids Chemistry Research Group under the guiding of Dr. S.Yarmoluk. Her present researches are devoted to the characterization and studies of mechanism of fluorescent cyanine dyes – biopolymers interaction with the use of spectral-luminescent methods [2].

I.V. Valyukh, V. B. Kovalska, Y. L. Slominskii, S. M. Yarmoluk (2002) Journal of Fluorescence *12*, 105-107.
T.Yu. Ogulchansky, M.Yu. Losytskyy, V.B. Kovalska, S.S. Lukashov, V.M. Yashchuk, S.M. Yarmoluk (2001) Spectrochim. Acta Part A: Mol. & Biomol. Spectroscopy, *57*, 2705-2715.

Date submitted: 11th September 2002

Ruud Kraayenhof, Ph.D.

Dept. of Structural Biology, Inst. of Molecular Cell Biology,
Vrije Universiteit, De Boelelaan 1085
1183 AZ Amsterdam,
The Netherlands.
Tel: +31 20 4447171
kr@bio.vu.nl

Specialty Keywords: Membranes, Protein dynamics .

Our research is focussed on (a) the structural dynamics of membrane proteins during catalytic action (e.g. ATP synthase), and (b) membrane surface properties, such as charge density, anisotropy, microviscosity and curvature, playing a role in the switching and regulation of protein functioning. Some new coumarin probes monitoring such membrane properties have been synthesized.

R. Kraayenhof, G.J. Sterk and H.W. Wong Fong Sang (1993) *Biochemistry* **32**, 10057–10066.
R.M. Epand, R. Cornell, S.M.A. Davies and R. Kraayenhof (2002) in R. Kraayenhof, A.J.W.G. Visser, and H.C. Gerritsen (Eds.), *Fluorescence Spectroscopy, Imaging and Probes*, Springer Series on Fluorescence, Vol. 2, Springer Verlag, Heidelberg, pp. 263–276.

Date submitted: 6th August 2002

Mikael Kubista, Ph.D.

TATAA Biocenter,
Medicinarg. 7B, Goteborg,
405 30,
Sweden.
Tel: +46 31 7733926   Fax: +46 31 7733948
Mikael.kubista@tataa.com
www.tataa.com

Specialty Keywords: Spectroscopy, Fluorescent probes, Realtime PCR.

Our research interest spans from characterization of molecular interactions by multidimensional spectroscopy (www.multid.se) to the development of fluorescent probes (www.lightup.se). Currently we design probes for real-time PCR applications and develop assays for accurate gene expression analysis in complex biological samples, including individual cells. We also develop Q-PCR assays for protein detection.

Eliminating the need for reference samples. Critical Reviews in Anal. Chem. 29, 1-28 (1999).
Light-up Probes. Anal. Biochem. 281, 26-35 (2000).

Date submitted: 28th August 2002

Alexander V. Kukhta, Ph.D.

Sector of Electron Spectroscopy and Optics,
Institute of Molecular and Atomic Physics,
National Academy of Sciences of Belarus,
F.Skaryna Ave. 70, Minsk 220072,
Belarus.
Tel: (+375) 17 2841719   Fax: (+375) 17 2840030
Kukhta@imaph.bas-net.by

Specialty Keywords: Organic electroluminescence, Electron-organic molecule interaction, Laser Physics.

Spectral and luminescent properties of interaction of low and high energy electrons with complex organic molecules having optoelectronic and biological interest, in the gas, liquid, and solid state. Organic electroluminescent materials and structures. Physics of energy transformation in organic electroluminescent structures. Excitation energy dependence of stimulated emission of organic materials under optical and electrical excitation.

S. M. Kazakov, A. V. Kukhta, and V. A. Suchkov (2000) *J. Fluorescence* **10,** 409-412.
A. V. Kukhta, G. G. Gorokh, E. E. Kolesnik, A. I. Mitkovets, M. I. Taoubi, Y. A. Koshin, and A. M. Mozalev (2002) *Surf. Sci.* **507-510**, 593-597.

Date submitted: 9th May 2002

Jens M. Kürner, Ph.D.

Competence Center for Fluorescent Bioanalytics
Josef-Engert-Str. 9
D – 93053 Regensburg, Germany.
Tel.: +49 (0) 941 943 5011   Fax: +49 (0) 941 943 5018
jens.kuerner@chemie.uni-regensburg.de
www.bioregio-regensburg.de/deutsch/partner/cfb.htm

Specialty Keywords: Array technolgy, Biotechnology, Microscopy / Flow-Cytometry, Synthesis / Spectroscopy.

Fluorescent bioanalytics is one of the most innovative research fields of modern life sciences with constantly increasing turnover rates. Within this market, the *Competence Center for Fluorescent Bioanalytics* is trying to establish itself as a competent service provider of customer-oriented research and development. Next to the diagnostical and research and development pharmaceutical industry, the competence center focuses on customers in biotechnology companies of the Regensburg region along with private and public research institutes.

The competence center has the objective to offer interdisciplinary research and development services in fluorescent bioanalytics in a unique network. This concept is based on the integration of components in chemistry, biology, medicine and engineering sciences along with the research and development potential of the University of Regensburg, the University of Applied Sciences of Regensburg and the University Hospital of Regensburg.

Date submitted: 31st August 2002

Akihiro Kusumi, D.Sc.

Department of Biological Science,
Nagoya University Chikusa-ku,
Nagoya, 464 8602,
Japan.
Tel: +81 52 789 2969   Fax: +81 52 789 2968
akusumi@bio.nagoya-u.ac.jp
www.supra.bio.nagoya-u.ac.jp

Specialty Keywords: Single molecules, Cell membrane, Signal transduction.

We develop single molecule techniques to be used for the study of live cells, such as single particle tracking and single fluorophore video imaging of membrane proteins, and single molecule dragging of membrane molecules using optical traps. Using these technologies, we study the mechanisms of signal transduction in the cell membrane, development of neuronal network, interaction of the membrane skeleton with membrane molecules, and formation and the functional mechanism of rafts, caveolae, and coated pits.

T. Fujiwara, K. Ritchie, K. Metz-Honda, K. Jacobson, and A. Kusumi. Phospholipids undergo hop diffusion in compartmentalised cell membrane. J. Cell Biol. 157, 1071-1081 (2002).
R. Iino, I. Koyama, and A. Kusumi. Single molecule imaging of GFP in living cells: E-cadherin forms oligomers on the free cell surface. Biophys. J. 80, 2667-2677 (2001).

Date submitted: 16th September 2002

Alexey S. Ladokhin, Ph.D.

University of California at Irvine,
Department of Physiology and Biophysics,
Irvine, CA 92697-4560,
USA.
Tel: (949) 824 6993   Fax: (949) 824 8540
ladokhin@uci.edu
krypton.biomol.uci.edu/ladokhin.html

Specialty Keywords: Membrane protein insertion, Depth-dependent quenching, Red-edge effects.

My research focuses on understanding the structural and thermodynamic principles of insertion and assembly of membrane proteins and uses fluorescence as a principal tool. Over the years we have developed and applied fluorescence methods enabling us to characterize the depth of membrane penetration into the bilayer, the lipid exposure and cis/trans topology of particular sites as well as the conformational heterogeneity of membrane-inserted proteins and peptides. Reprints are available at: http://blanco.biomol.uci.edu/reprints/index.html.

A. S. Ladokhin (1999). Analysis of protein and peptide penetration into membranes by depth-dependent fluorescence quenching: Theoretical considerations. *Biophys. J.* 76:946-955.
A. S. Ladokhin, S. Jayasinghe and S. H. White (2000). How to measure and analyze tryptophan fluorescence in membranes properly, and why bother? *Anal. Biochem.* 285:235-245.

Date submitted: 6th May 2002

Joseph R. Lakowicz, Ph.D.

Center for Fluorescence pectroscopy,
Dept. of Biochemistry and Molecular Biology,
University of Maryland School of Medicine,
725 West Lombard St, Baltimore, Maryland, 21201, USA.
Tel: 410 706 7978   Fax: 410 706 8409
Lakowicz@cfs.umbi.umd.edu
cfs.umbi.umd.edu

Specialty Keywords: Fluorescence.

Current Research Interests: My research is focused on advancing the field of fluorescence spectroscopy. This involves chemical synthesis of new fluorophores, development of novel fluorescence measurements, development of instrumentation for time-resolved fluorescence, and the chemical applications of fluorescence sensing. More recently I have been investigating the effects of metallic surfaces on modifying the *free-space* fluorescence spectral properties of near-by fluorophores, with potential applications spanning the analytical and biomedical sciences.

J. R. Lakowicz (2001). Radiative Decay Engineering: Biophysical and Biomedical Applications, *Anal. BioChem.,* **298**, 1-24.

J. R. Lakowicz, I. Gryczynski, Y. B. Shen, J. Malicka, and Z. Gryczynski, (2001). Intensified fluorescence, *Photonics Spectra,* **35**(10), 96-104.

Date submitted: 24th August 2002

Marek J. Langner, Ph.D.

Department of Physics,
Wrocław University of Technology,
Wyb. Wyspiańskiego 27, Wrocław,
PL – 50-370, Poland.
Tel: 48 (71) 320 23 84   Fax: 48 (71) 328 36 96
langner@rainbow.if.pwr.wroc.pl

Specialty Keywords: Supramolecular aggregates, Biosensors, Liposomes.

Applying fluorescence techniques to constructing, validating and determining properties of supramolecular aggregates including liposome based biosensors, lipoplexes and particulate drug carriers. Current research includes DNA condensation, surface electrostatics, aggregate topology, lipoplex association with cells and intracellular distribution. Fluorescence techniques used: fluorescence spectroscopy, FCS, fluorescence microscopy, FACS.

S. W Hui, M. Langner, Y. L. Zhao, P. Ross, E. Hurley and K. Chan (1996) Biophys. J. 71, 590-599.

T. Kral, M. Langner, M. Benes, D. Baczynska and M. Hof (2002) Bioph. Chem. 95, 135-144

Date submitted: 30th August 2002

Thomas M. Laue, Ph.D.

Biochemistry and Molecular Biology,
U. New Hampshire Rudman Hall, 46 College Road,
Durham, NH 03824, USA.
(Center to Advance Molecular Interaction Science (CAMIS))
(Biomolecular Interaction Technologies Center (BITC))
Tel: (603) 862 2459   Fax: (603) 862 0031
tom.laue@unh.edu
www.camis.unh.edu & www.bitc.unh.edu
Specialty Keywords: Fluorescence optics, Analytical
ultracentrifuge, Binding strength and Characterization.

CAMIS develops unique instruments to characterize molecular interactions such as a fluorescence detector for the AUF. BITC is an NSF Industry/University Cooperative Research Center composed of global pharmaceutical firms and instrument manufacturers.

Laue, T.M. and Stafford, W.F. III (1999) "Modern Applications of Analytical Ultracentrifugation," Annual Review of Biophysics and Biomolecular Structure V. 28, 75-100.
Laue, T.M., Anderson, A.L. and Weber, B.W. (1997) "Prototype Fluorescence Detector for the XLA Analytical Ultracentrifuge" in Ultrasensitive Clinical Laboratory Diagnostics, SPIE Proceedings, V. 2985, pp. 196-204, G. Cohn and S. Soper eds., SPIE, Bellingham, WA.

Date submitted: 13th September 2002

Robert P. Learmonth, Ph.D.

Department of Biological and Physical Sciences,
University of Southern Queensland,
Toowoomba QLD 4350,
Australia.
Tel: 61 7 4631 2361   Fax: 61 7 4631 1530
learmont@usq.edu.au
www.usq.edu.au/biophysci
Specialty Keywords: Multi-photon microscopy, Yeast, Membrane
fluidity.

Research areas: yeast biotechnology, cell membrane biochemistry/biophysics. Using yeast as a model system to investigate how cells react to changes in environment, focusing on cell membranes as the critically important structures in adaptation. Development of methods using novel fluorescent probes and multi-photon microscopy to study membrane status in single cells of yeasts, bacteria and other microbes.

Learmonth, R.P. and Gratton, E. Assessment of Membrane Fluidity in Individual Yeast Cells by Laurdan Generalized Polarization and Multi-Photon Scanning Fluorescence Microscopy. In Fluorescence Spectroscopy, Imaging and Probes - New Tools in Chemical, Physical and Life Sciences (R Kraayenhof, AJWG Visser and HC Gerritsen, Eds.), Springer Series on Fluorescence: Methods and Applications, Vol. 2, Springer, Heidelberg, 2002 (In Press), Chapter 14, ISBN 3-540-42768-6.

Date submitted: 8th July 2002

W. Jonathan Lederer, M.D., Ph.D.

Medical Biotechnology Center,
University of Maryland Biotechnology Institute,
725 W. Lombard Street, Baltimore, MD
USA 21201.
Tel: 410 706 8181   Fax: 410 706 8184
lederer@umbi.umd.edu
www.umbi.umd.edu/~mbc/pages/lederer.htm

Specialty Keywords: Heart, Confocal microscopy, Patch clamp, Calcium.

Work in the lab focuses on Ca^{2+} signaling in cardiac and other living cells. By combining confocal, multiphoton or wide-field microscopy with whole cell patch clamp techniques, we have been able to investigate the effects of subcellular and intracellular Ca^{2+} concentration ($[Ca^{2+}]i$ on cellular function. Diverse additional tools are used as needed including flash photolysis of caged chemicals, multi-photon uncaging, single channel examination in planar lipid bilayers and by patch clamp, immuno-fluorescence imaging, use of cells from transgenic and gene knockout animals, and use of primary cultures and co-cultures. . Much of the recent work focuses on "calcium sparks" and how the heart works in health and disease.

Nelson, M.T., Cheng, H., Rubart, M., Santana, L.F., Bonev, A., Knot, H. & Lederer, W.J. (1995). Relaxation of arterial smooth muscle by calcium sparks. Science 270:633-637.

Date submitted: 3rd September 2002

Thomas S. Lee, M.Sc.

International School of Photonics,
Cochin University of Science and Technology,
Kalamassery, Cochin, India 682 022
India.
Tel: +91 484 575848   Fax: +91 484 576714
lee@cusat.ac.in
www.photonics.cusat.edu

Specialty Keywords: Sol-gel, pH sensors, Fiber optic sensors.

I have carried out extensive research in the development of fiber optic sensors for chemical and physical applications. Chemical sensors include pH sensors based on dye impregnated sol-gel coatings. Also I have prepared bulk dye doped xerogels for quantum yield measurements, thermal lens spectroscopy and nonlinear applications in collaboration with other scientists.

Thomas Lee S, B Aneeshkumar, P Radhakrishnan, C P G Vallabhan and V P N Nampoori, *A microbent fiber optic pH sensor,* Opt. Comm **205**, 253 – 256 (2002)
Thomas Lee S, Nibu A George, P Sureshkumar, P Radhakrishnan, C P G Vallabhan and V P N Nampoori, *Chemical sensing with microbent optical fiber*, Opt. Lett., **20**, 1541-1543 (2001)

Date submitted: 21ˢᵗ August 2002

Barry R. Lentz, Ph.D.

Departmnt of Biochemistry & Biophysics CB#7260,
Molecular and Cellular Biophysics,
University of North Carolina at CH,
Chapel Hill, NC 27599-7260, USA.
Tel: 919 966 5384   Fax: 919 966 2852
uncbrl@med.unc.edu
hekto.med.unc.edu:8080/FACULTY/LENTZ/lab.html

Specialty Keywords: Membrane Probes, Phase Fluorescence, Fusion Assays.

Dr. Lentz's lab uses fluorescence spectroscopy to examine protein-lipid interaction involved in prothrombin activation during blood coagulation. The lab has shown that specific sites on blood coagulation proteins recognize phosphatidylserine, and that this lipid, which is exposed during platelet activation, regulates these proteins. Lentz's lab is also a leader in the application of fluorescence methods to studying the kinetics of lipid rearrangements during membrane fusion. Using these methods, the Lentz lab has developed a model for the mechaism of fusion as it occurs in model membranes and may well occur in biological membranes during such processes as viral infection and neurotransmitter release.

Date submitted: 22ⁿᵈ August 2002

Panagiotis Lianos, Ph.D.

University of Patras,
Engineering Science Dept,
26500 Patras,
Greece.
Tel: 30 610 997587   Fax: 30 610 997803
lianos@upatras.gr
www.des.upatras.gr

Specialty Keywords: Applied Photophysics and Photochemistry.

Recent research focuses on the study of photophysical and photochemical applications of nanocomposite organic/inorganic materials made by soft chemistry procedures (sol-gel method). Applications include dye-sensitized photoelectrochemical cells, photocatalytic metal oxide surfaces and new photoluminescence and electroluminescence light sources based on ligand-lanthanide ion complexes.

E. Stathatos, P.Lianos and Ch.Krontiras (2001) *J.Phys.Chem. B.*   105, 3486-3492
V.Bekiari and P.Lianos (2000) *Adv.Mater.* 12, 1603-1605.

Date submitted: 6th August 2002

David M. J. Lilley, FRS

MSI / WTB Complex,
University of Dundee,
Dundee, DD1 5EH,
United Kingdom.
Tel: +44 1382 344243   Fax: +44 1382 345893 / 201063
d.m.j.lilley@dundee.ac.uk
www.dundee.ac.uk/biocentre/nasg/

Specialty Keywords: FRET, Nucleic acid structure.

Our interests are directed at the structure and folding of branched nucleic acids; the four–way junction in DNA, and a variery of structures (especially ribozymes) in RNA. Our main biophysical approach is fluorescence resonance energy transfer (FRET), in steady state, time-resolved and single-molecule modes.

D.A. Lafontaine, D.G. Norman and D. M.J. Lilley. The global structure of the VS ribozyme. EMBO J. 21, 2461-2471 (2002).

T.J. Wilson and D.M.J. Lilley Metal ion binding and the folding of the hairpin ribozyme RNA 8, 587-600 (2002).

Date submitted: 26th August 2002

M. Pilar Lillo, Ph.D.

Instituto Química Física "Rocasolano", C.S.I.C.,
Serrano 119, 28006 Madrid,
Spain.
Tel: 34 91 5619400 ext 1400   Fax: 34 91 5642431
pilar.lillo@iqfr.csic.es

Specialty Keywords: Time-resolved fluorescence, FRET, Biomolecular interactions.

Current interest: Design of fluorescence anisotropy and FRET methodologies for ligand binding, protein-protein, and protein-DNA interaction studies.
Structural and dynamical characterization of symmetrical homopolymers by Förster resonance energy homo transfer (FREHT).

M.P. Lillo, O. Cañadas, R.E.Dale, A.U.Acuña (2002). The location and properties of the taxol binding center in microtubules: a ps laser study with fluorescent taxoids. *Biochemistry (in press)*
M.P. Lillo, B.K.Szpikowska, M.T.Mas, J.D. Sutin and J.M.Beechem (1997). Real-time measurement of multiple intramolecular distances during protein folding reactions: a multisite stopped-flow fluorescence energy-transfer study of PGK. *Biochemistry* 36, 11273-11281

Date submitted: 31st August 2002

Marcin Lipski, Ph.D.

Poznan University of Technology,
Institute of Chemistry & Technical Electrochemistry,
Piotrowo 3, 60 965 Poznań,
Poland.
Tel: +48 (61) 665 2068   Fax: +48 (61) 665 2571
mlipski@sol.put.poznan.pl
www.put.poznan.pl

Specialty Keywords: Photochemistry & Molecular Spectroscopy of Humic Acids & Precursors-Hydroxybenzotropolones.

Current Research Interests: Fluorescence of humic acids and unusual precursors - purpurogallin (2,3,4,6–tetrahydroxy–5H-benzocyclohepten–5–one, hydroxybenzotropolone) and its analogues formed from the polyphenols.

M. Lipski (2002). Fluorescence emitted during the autooxidation of 2,3,4,6-tetrahydroxy-5H-benzocyclohepten-5-one, *Journal of Fluorescence,* **12**(1), 83-86.
M. Lipski, K. Gwozdzinski, J. Slawinski (2000). Free radical of the semiquinone type generated in the redox reaction of hydroxybenzotropolone, *Current Topics in Biophysics,* **24**(2), 115-120.
M. Lipski, J. Slawinski, D. Zych (1999). Changes in the luminescent properties of humic acids induced by UV-radiation, *Journal of Fluorescence,* **9**(2), 133-138.

Date submitted: 30th August 2002

Burton J. Litman, Ph.D.

Section of Fluorescence Studies,
Laboratory of Membrane Biochemistry and Biophysics,
National Institute on Alcohol and Alcoholism,
National Institutes of Health, 12420 Parklawn Drive, Rm 114,
Rockville, Maryland, 20852.
Tel: 301 594 3608   Fax: 301 594 0035
litman@helix.nih.gov

Specialty Keywords: Membrane structure, Fluorescent probes, GPCR signaling systems.

Research interests focus on the effect of lipid composition on GPCR signaling, using the visual transduction system as a model. The role of polyunsatrurated phospholipids and cholesterol in modulating signaling and domain formation is investigated. Membrane phospholipid acyl chain packing and domain formation are monitored using various fluorescence techniques.

S-L Niu, D. C. Mitchell, and B. J. Litman (2002) Manipulation of Cholesterol Levels in Rod Disk Membranes by Methyl-β-cyclodextrin. Effects On Receptor Activation, J. Biol. Chem. **277**: 20139-20145.
A. Polozova and B. J. Litman (2000) Cholesterol Dependent Recruitment of di22:6-PC by a G Protein-Coupled Receptor into Lateral Domains, Biophys. J. **79**, 2632–2643.

Date submitted: 3rd September 2002

Garrick M. Little, Ph.D.

Li-Cor,
4308 Progressive Ave, Lincoln,
Nebraska, 68504,
USA.
Tel: (402) 467 0716   Fax: (402) 467 0819
glittle@licor.com
WWW.licor.com

Specialty Keywords: Protein labeling, Western blot assay, DNA labeling.

My research interests include the synthesis of Infra-red fluorescent dyes functionalized as the amidite, NHS ester etc. Labeling of biological molecules with fluorescent dyes. More generally Organic chemistry synthesis, synthesis of DNA.

Date submitted: 4th September 2002

David Lloyd, Ph.D., D.Sc.

Microbiology (BIOSI 1, Main Bldg),
Cardiff University,
P.O. Box 915,
Wales, U.K.
Tel: 44 (0)29 2087 4772   Fax: 44 (0)29 2087 4305
lloydd@cardiff.ac.uk

Specialty Keywords: Bioenergetics, Mitochondria, Oscillations.

Mitochondrial inner membrane electrochemical potential measurements have been used to investigate the respiratory oscillations that indicate the operation of the ultradian clock in yeasts and protists. Plasma membrane potential measurements quantify perturbation of organisms by inhibitors, biocides and antibiotics to provide an indication of vitality (e.g. for fermentative efficiency in yeast inocula for commercial processes) or loss of viability (e.g. as an indicator of antibiotic sensitivity).

Lloyd, D. *et al*. (2002) Cycles of mitochondrial energization driven by the ultradian clock in continuous culture of *Saccharomyces cervisiae*. *Microbiology* **148**, in press.

Suller, M.T.E. and Lloyd, D. (2002) The antibacterial activity of Valinomycin towards *Staphylococcus aureus* under aerobic and anaerobic conditions. *J. Appl. Bact*. **92**, 866-72.

Date submitted: 26ᵗʰ August 2002

Leslie M. Loew, Ph.D.

Center for Biomedical Imaging Technology,
University of Connecticut Health Center,
Farmington, CT 06030 1507,
USA.
Tel: 860 679 3568   Fax: 860 679 1039
les@volt.uchc.edu
www.cbit.uchc.edu/

Specialty Keywords: Non-linear optical microscopy, Dye synthesis, Cell physiology.

We have a long-standing effort on the synthesis of voltage-sensitive dyes which has recently led us to develop dyes and optical systems for second harmonic imaging microscopy. We have also been developing a computational system called "Virtual Cell" for modeling and simulating cellular events based on microscope images. Our biological research focuses on mapping the electrical profiles along cell surfaces and exploring their cell physiological implications.

Slepchenko B, Schaff JC, Carson JH, Loew LM. 2002. Computational cell biology: spatiotemporal simulation of cellular events. Annual Review of Biophysics & Biomolecular Structure 31:423-441.

Campagnola, P. J., H. A. Clark, W. A. Mohler, A. Lewis, and L. M. Loew. 2001. Second Harmonic Imaging Microscopy of Living Cells, J. Biomedical Optics, 6:277-286.

Date submitted: 5ᵗʰ September 2002

Piet H.M. Lommerse

Dept. of Biophysics, Leiden University,
Niels Bohrweg 2, Leiden,
2333 CA,
The Netherlands.
Tel: (+31) 71 527 5946   Fax: (+31) 71 527 5819
plommerse@biophys.leidenuniv.nl
www.biophys.leidenuniv.nl/Research/FvL

Specialty Keywords: Single-molecule, Fluorescence, Microscopy.

In the last decade evidence has accumulated that small domains (26-700 nm diameter) are located in the plasma membrane. Using wide-field fluorescence microscopy with single-molecule sensitivity, the diffusion of individual membrane anchored eYFP molecules is studied in live cells at the millisecond timescale, to reveal the intricate details of membrane organization and its role in signal transduction.

Gregory S. Harms, Laurent Cognet, Piet H.M. Lommerse, Gerhard A. Blab and Thomas Schmidt (2001) *Biophysical Journal*, **80**, 2396-2408.

Lopez, A.
Losytskyy, M.U.

Date submitted: 28th August 2002

André Lopez, Ph.D.

Institut de Pharmacologie et de Biologie Structurale du CNRS,
205 route de Narbonne,
Toulouse, 31400,
France.
Tel: (33) 561 175 945   Fax: (33) 561 175 994
andre.lopez@ipbs.fr
ipbs.fr

Specialty Keywords: Membrane probes, Multichromophoric systems, Biomembranes.

Functional consequences of membrane composition and microcompartmentation in connection with the translational dynamics of lipids and proteins on the chain of signal transduction by G protein-coupled receptors. These studies are carried out on human receptors µ, CCR5, CXCR4 expressed in various cell types. Are investigated: (i) the influence of lipid environmental factors on receptor activity, (ii) the lateral dynamics and compartimentations of these membrane compounds using fluorescence techniques (FRAP, SPT), (iii) the structure *in situ* of these pluri-molecular systems by means of spectromicrofluorescence approaches (FRET, polarity probes).

Date submitted: 30th August 2002

Mykhaylo Yu. Losytskyy, M.Sc.

Institute of Molecular Biology and Genetics of NAS of Ukraine,
Zabolotnogo Str. 150, Kyiv,
03143, Ukraine.
Tel: (+38 044) 252 23 89   Fax: (+38 044) 252 24 58
m_losytskyy@svitonline.com
www.yarmoluk.org.ua

Specialty Keywords: Energy transfer, J-aggregate, Cyanine dye.

The studies of M. Losytskyy are aimed on the designing of fluorescent probes for nucleic acid and protein detection. Now he is working in Nucleic Acids Chemistry Research Group under the guiding of Dr. S.Yarmoluk. His present studies are devoted to the spectroscopy study of electronic excitation energy transfer in DNA-cyanine dye system; photophysics of the excited dye molecules [1]; and J-aggregates of cyanine dye formed on nucleic acids [2].

M. Yu. Losytskyy, V. M. Yashchuk, S. S. Lukashov, S. M. Yarmoluk (2002) *Journal of Fluorescence* 12, 109-112.
S.M. Yarmoluk, M.Yu. Losytskyy, V.M. Yashchuk (2002) *J. Photochem. Photobiol.* 67, 57-63.

Date submitted: 11th September 2002   **Luís M. S. Loura, Ph.D.**

Centro de Química-Física Molecular, IST,
Av. Rovisco Pais,
1049-001 Lisbon,
Portugal.
Tel: +351 21 8419219 Fax: +351 21 8464455
pclloura@alfa.ist.utl.pt

Specialty Keywords: FRET, Lipid domains, Lipid protein-interaction.

Current Research Interests: Study of membrane heterogeneities (domains/rafts) using photophysical methodologies. Derivation of kinetic models for FRET in restricted geometries. Development of software for global analysis of fluorescence decays. Topology and dynamics of protein/peptide interaction with model systems of membranes. Cholesterol organization in membranes. Characterization of DNA/cationic lipid complexes.

L. M. S. Loura, A. Fedorov and M. Prieto (2001) *Biophys. J.* **80**, 776-788.
L. M. S. Loura, R. F. M. de Almeida and M. Prieto (2001) *J. Fluorescence* **11**, 197-209.

Date submitted: 12th September 2002   **Joanna Malicka, Ph.D.**

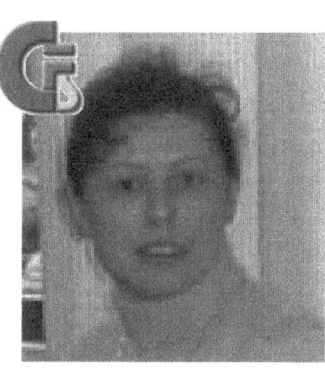

Center for Fluorescence Spectroscopy,
Medical Biotechnology Center,
University of Maryland, 725 West Lombard St,
Baltimore, Maryland, 21201, USA.
Tel: 410 706 7500 Fax: 410 706 8409
malicka@umbi.umd.edu
cfs.umbi.umd.edu.

Specialty Keywords: FRET, Metal-enhanced fluorescence, Conformational analysis of peptides.

My expertise includes multi-step synthesis of peptides and fluorophore derivatives of amino acids as well as a conformational analysis of bioactive peptides by using NMR spectroscopy and FRET. I have experience in steady-state and time-resolved fluorescence measurements and in metallic colloids and surface preparation. My current interest is focused on metal-fluorophore interactions in solution and on surfaces and their application to a new generation of very efficient biological assays based on enhanced-fluorescence and changes of FRET efficiency near silver particles.

Radiative decay engineering. 2. Effects of silver island films on fluorescence intensity, lifetimes, and resonance energy transfer (2002). Lakowicz J.R., Shen Y., D'Auria S., Malicka J., Fang J., Gryczynski Z., Gryczynski I., Anal. Biochem., 301, 261-2.

Date submitted: 1st September 2002

Emmanuel Margeat, Ph.D.

Single Molecule Biophysics Group,
UCLA Departement of Chemistry and Biochemistry,
607 Charles E Young Drive East, Los Angeles,
CA 90095, USA.
Tel: (1) 310 794 6693
margeat@chem.ucla.edu
smb.chem.ucla.edu

Specialty Keywords: FRET, Polarization, Single molecule.

My objective is to elucidate the structure and dynamics of protein / protein and protein / nucleic acids complexes using a combination of novel single-molecule fluorescence microscopy methods (such as spFRET, fluorescence anisotropy) and traditional biochemistry. My research focuses on macromolecular complexes involved in transcription, including nuclear receptors, coactivators, and RNA polymerase.

Margeat E., Poujol N., Boulahtouf A., Chen Y., Gratton E., Cavaillès V. and Royer C. "The Estrogen Receptor binds a single SRC-1 coactivator molecule with an affinity dictated by the agonist structure." *Journal of Molecular Biology,* 306 (3):433-442 (2001).

Date submitted: 5th September 2002

Mark Maroncelli, Ph.D.

Department of Chemistry, Penn State University,
152 Davey Laboratory, University Park,
PA, 16802,
USA.
Tel: 814 863 5319
mpm@chem..psu.edu
maroncelli.chem.psu.edu

Specialty Keywords: Time-resolved fluorescence, Ultrafast spectroscopy, Solution dynamics.

We use steady-state and ultrafast fluorescence spectroscopy and computer simulations to explore solvation and its influence over chemical processes in liquid solvents and supercritical fluids.

J. Lewis, R. Biswas, A. Robinson, and M. Maroncelli (2001)., Local Density Augmentation in Supercritical Fluids: Electronic Shifts of Anthracene Derivatives *J. Phys. Chem. B* **105**, 3306.

M. L. Horng, J. A. Gardecki, A. Papazyan, and M. Maroncelli (1995)., Sub-Picosecond Measurements of Polar Solvation Dynamics: Coumarin 153 Revisited *J. Phys. Chem.* **99**, 17311.

Date submitted: 12st September 2002

José M. G. Martinho, Ph.D.

Centro de Química-Física Molecular,
Instituto Superior Técnico,
1049-001 Lisboa,
Portugal.
Tel: +351 218419250   Fax: +351 218464455
jgmartinho@ist.utl.pt

Specialty Keywords: Photophysical kinetics, Resonance energy transfer, Polymers, Colloids.

Current interests: Conformation and dynamics of proteins and oligonucleotides adsorbed onto latex particles. Phtophysical kinetics (early work included the study of transient effects in pyrene monomer–excimer kinetics). Radiative transport in scattering media. Conformations and aggregation of polymers in solutions. Interfaces in polymer systems.

J. M. G. Martinho, J. P. Farinha, M. N. Berberan-Santos, J. Duhamel, M. A. Winnik (1992).Test of a model for reversible excimer kinetics: Pyrene in cyclohexanol, *J. Chem. Phys.* **96**, 8143.
S. Piçarra, J. M. G. Martinho (2001). Viscoelastic effects on dilute polymer solutions phase demixing: Fluorescence study of a poly(m-caprolactone) chain in THF, *Macromolecules* **34**, 53.

Date submitted: 3rd September 2002

Masayuki Masuko, Ph.D.

Hamamatsu Photonics K. K., Tsukuba Research Laboratory,
5-9-2 Tokodai, Tsukuba,
300-2635,
Japan.
Tel: +81 298 47 5161   Fax: +81 298 47 5266
masuko@hpk.trc-net.co.jp
www.hpk.co.jp/

Specialty Keywords: Nucleic acids, Excimer fluorescence, Photon counting.

I am interested in the application of aromatic hydrocarbon dyes to the detection of biological substances such as nucleic acids, and the development of instruments useful to their measurements.

M. Masuko, H. Ohtani, K. Ebata and A. Shimadzu (1998) Optimization of excimer-forming two-probe nucleic acid hybridization method with pyrene as a fluorophore *Nucleic Acids Res.* **26** (23), 5409-5416.
M. Masuko, S. Ohuchi, K. Sode, H. Ohtani and A. Shimadzu (2000) Fluorescence resonance energy transfer from pyrene to perylene labels for nucleic acid hybridization assays under homogeneous solution conditions *Nucleic Acids Res.* **28** (8), e34.

Mathis, G.
Matkó, J.

Date submitted: 17th September 2002 **Gerard Mathis, D.Sc.**

CIS biointernational,
BP84175 Bagnols sur Ceze,
F 30204,
France.
Tel: 33 (0) 4 66 79 67 71   Fax: 33 (0) 4 66 79 19 20
gmathis@cisbiointernational.fr

Specialty Keywords: Rare earth cryptates synthesis and fluorescence, FRET, Biomolecular interactions.

Current interests: Design of luminescent rare earth cryptates and photophysical studies. Research of fluorescence based techniques for probing interactions between biomolecules. Research and development of methods based on the use of long lived fluorophores and Fluorescence resonance energy transfer. Applications in cellular and molecular biology.

H.Bazin,E.Trinquet,G.Mathis (2002). Time Resolved Amplification of Cryptate Emission: a Versatile Technology to Trace Biomolecular Interactions. Reviews in Molecular Biotechnology **82,, 233-250.**

Date submitted: 21st August 2002 **János Matkó, Ph.D., D.Sc.**

Department of Immunology, Eotvos Lorand University,
Pázmány Péter sétány 1/C, Budapest,
H-1117,
Hungary.
Tel: (36) 1 3812175 Fax: (36) 1 3812176
matko@cerberus.elte.h

Specialty Keywords: Microscopy, Flow cytometry, FRET.

Developing/applying fluorescence techniques (e.g. FRET, FPR, polarization) in studies of supramolecular organization/mobility of receptor and signals proteins, as well as lipid rafts at the surface (plasma membrane) of immunecompetent cells, investigations on the functional role of protein clustering.
Techniques: Fluorescence Resonance Energy Transfer w. flow cytometry/pbFRET microscopy; confocal microscopy, SNOM, FPR, time-resolved phosphorescence spectroscopy).

Matkó J., Edidin, M., *Methods in Enzymology*, Vol.278, 444-462 ,1997.
Vereb, G., Matkó, J., et al., *Proc. Natl. Acad. Sci. USA*, 97, 6013-6018, 2000

Date submitted: 9th September 2002

James R. Mattheis, Ph.D.

SPEX Fluorescence, Jobin Yvon, Inc.,
3880 Park Ave, Edison,
NJ, 08820-3012,
USA.
Tel: 732 494 8660 ext: 122
Jim_mattheis@jyhoriba.com

Specialty Keywords: Photon-counting, Frequency-domain.

Managing a team of scientists providing fluorescence applications support, training and new methods development for users of SPEX spectrofluorometers. Support is provided for all users interested in applying high sensitivity photon-counting, steady-state fluorescence spectroscopy, fluorescence microscopy and picosecond time-resolved, frequency-domain methods to their own research projects.

Date submitted: 6th September 2002

Evgenia G. Matveeva, Ph.D.

Burstein Technologies, Inc.,
163 Technology Drive, Irvine,
CA, 92618,
USA.
Tel: (949) 453 1800   Fax: (949) 453 1818
ematveeva@bursteintech.com

Specialty Keywords: Immunoassays, Fluorescence.

Homogeneous immunoassay based on reverse micellar systems, quenching and polarization fluoroimmunoassay in organic and aqueous media. Detection of paraoxone, atrazine, propazine, dioxin, pyrethroids, steroids, porphyrins, proteins. Tube-format and plate-format assays. Compact-disc-format heterogeneous competitive immunoassays for cardiac markers and pregnancy marker.

Matveeva E.G., Shan G., Kennedy I.M., Gee S.J., Stoutamire D.W., and Hammock B.D. (2001): Homogeneous Fluoroimmunoassay of a Pyrethroid Metabolite in Urine. *Analitica Chim. Acta*, v.444, 103-117.

Matveeva E.G., Nelen M.I., Lobanov O.I., and Savitsky A.P. Specific Interaction of Coproporphyrin I with Antibodies in Water and Reverse Micelles. Dimerization of antibodies affects antigen biding site. *J. Fluorescence*, 2003, accepted.

Date submitted: 7th July 2002

László Mátyus, M.D., Ph.D.

Department of Biophysics and Cell Biology,
University of Debrecen,
Nagyerdei krt 98, Debrecen,
H-4012, Hungary.
Tel / Fax: +36 52 412 623
lmatyus@jaguar.dote.hu

Specialty Keywords: Fluorescence resonance energy transfer.

My research interest is to study the distribution and conformation of cell surface receptors using various fluorescence techniques, such as flow cytometric energy transfer measurements or different microscopies.

L. Mátyus, L. Bene, J. Hársfalvi,, M.V. Alvarez, J. González-Rodríguez, A. Jenei, L. Muszbek, and S. Damjanovich, (2001). Organization of the glycoprotein (GP) IIb/IIIa heterodimer on resting human platelets studied by flow cytometric energy transfer *J. Photochem. Photobiol. B: Biol.* **65** 47-58.

P. Nagy, L. Mátyus, A. Jenei, G. Panyi, S. Varga, J. Matkó, J. Szöllősi, R. Gáspár, T.M. Jovin, and Damjanovich (2001). Cell fusion experiments reveal distinctly different association characteristics of cell surface receptors *J. Cell. Sci.* **114** 4063-4071.

Date submitted: 22nd August 2002

Vladimir M. Mazhul, Ph.D.

Laboratory of Protein Photonics, Institute of Photobiology,
of National Academy of Sciences of Belarus,
Akademicheskaya, 27, Minsk,
Belarus, 220072.
Tel: 375 17 2842251 Fax: 375 17 2842359
Ipb@biobel.bas-net.by

Specialty Keywords: Room temperature phosphorescence.

Specialist in the fields of studying proteins and lipid peroxidation (LPO) products by fluorescence and room temperature phosphorescence techniques. The systematic investigations of millisecond internal dynamics of proteins in solution and composition of cell membrane by room temperature tryptophan phosphorescence technique had been carried out. By room temperature phosphorescence method the heterogeneity of LPO products accumulation in bulk and annular lipids of the cellular membrane has been shown.

V.M. Mazhul', E.M. Zaitseva and D.G. Shcharbin (2000) *Biophysics* **45**, pp. 965-989.
V.M. Mazhul' and D.G. Shcharbin (2000) *Current Topics in Biophysics* **24** pp. 139-146.

Date submitted: 26th August 2002

Alberto Mazzini, Ph.D.

Department of Physics, University of Parma,
Parco Area Scienze 7A,
Parma, 43100,
Italy.
Tel: +39 0521 906229   Fax: +39 0521 905223
mazzini@fis.unipr.it

Specialty Keywords: Protein folding, Binding analysis of probes to proteins, Time Correlated Single Photon Counting.

My present research interest is to study denaturation and renaturation mechanisms of proteins. Unfolding is induced by chemical denaturants and refolding is achieved by recovery of native experimental conditions. Intrinsic and extrinsic fluorescence is studied both by stationary and time resolved techniques (TCSPC). In the case of simple monomeric or dimeric proteins such as odorant binding proteins (OBP), the thermodynamic and kinetic analysis allows to elucidate the unfolding/refolding mechanism.

A.Mazzini, A.Maia, M.Parisi, R.T.Sorbi, R.Ramoni, S.Grolli, R.Favilla (2002) *Biochim.Biophis Acta* 1599, 82-93.

R.Favilla, M.Goldoni, A.Mazzini, P. Di Muro, B.Salvato, M.Beltramini (2002) *Biochim.Biophis Acta* 1597, 42-50.

Date submitted: 30th August 2002

Claudia Mazzuca, (Ph.D. Student)

Department of Chemical Sciences and Technologies,
University of Roma Tor Vergata,
Via della ricerca scientifica, 00133, Roma,
Italy.
Tel: +39 06 7259 4469   Fax: +39 06 7259 4328
sopwithcamel76@hotmail.com

Specialty Keywords: Peptide structure, Foldamers, Peptide-membrane interactions.

My research activity within the group of professor Pispisa B. is focused on the use of fluorescence spectroscopy to investigate the interaction of antibiotic peptides with membranes and their mode of action.

I am interested also in determining the structure of synthetic, unusual amino acid based oligopeptide as foldamers.

B. Pispisa et al. (2000) Structural features of linear (☺Me)Val-based peptides in solution by photophysical and theoretical conformational studies. *Biopolymers* **55**, 425-435.

B. Pispisa et al. (2002) Effect of distortions on the optical properties of Amide NH Infrared Absorption in short peptide in solution. *J. Phys. Chem B* **106**, 5733-5738.

Date submitted: 22nd August 2002

Weiping Mei, Ph.D.

Head of Biophotonics,
R&D cosmed, Beiersdorf AG,
Unnastr.48, 22457 Hamburg,
Germany.
Tel: +49 40 4909 4896   Fax: +49 40 4909 18 4896
weiping.mei@beiersdorf.com
www.beiersdorf.com; www.nivea.de

Specialty Keywords: Biophotonics, Ultraweak photon emission, Chemiluminescence.

Research and methods development on the basis of detecting photons from human skin directly *in vivo*. The most interests is on the application of using optical technique for understanding photophysics and photochemistry of skin and efficacy test of skin care product.

W.P. Mei (1994) About the Nature of Biophotons. *Journal of Biological Systems*, Vol. 2, 25-42.

Sauermann G., Mei W.P., Hoppe U. and Stäb F.: Ultraweak Photon Emission of Human Skin *in vivo* - Influence of topically applied antioxidants on human skins. *Oxidants & Antioxidants, Part B, Methods in Enzymology*, Volume 300 (1999), p 419-428.

Date submitted: 8th July 2002

Yves Mely, Ph.D.

Université Louis Pasteur, UMR 7034 CNRS,
Faculté de Pharmacie, 74 route du Rhin,
67401 Illkirch,
France.
Tel: 33 (0) 3 90 24 42 63   Fax: 33 (0) 3 90 24 42 13
mely@pharma.u-strasbg.fr
umr7034.u-strasbg.fr/

Specialty Keywords: Time-resolved fluorescence, Fluorescence correlation spectroscopy, Protein interactions.

My research is mainly focused on the investigation by fluorescence techniques of the interaction of proteins (HIV nucleocapsid protein, elastase) with various ligands (ions, nucleic acids, peptides). I also investigate the physico-chemical properties and the intracellular fate of complexes of DNA with nonviral vectors. More recently, I have developed a platform with TPE that combines FCS, time-resolved fluorescence, microspectrofluorimetry and imaging.

S. Bernacchi, S. Stoylov, E. Piémont, D. Ficheux, B.P. Roques, J.L. Darlix & Y. Mély (2002). HIV-1 nucleocapsid protein activates transient melting of least stable parts of the secondary structure of TAR and its complementary sequence. *J. Mol. Biol.*, **317**, 385-399.

E. Bombarda, A. Ababou, D. Gérard, B.P. Roques, E. Piémont & Y. Mély (1999). Time-resolved fluorescence investigation of the HIV-1 nucleocapsid protein. *Biophys. J.*, **76**, 1561-1570.

Date submitted: 4th September 2002   **Francisco Mendicuti, Ph.D.**

Química Física, Univesidad de Alcalá,
Ctra Madrid-Barcelona Km 33.6,
28871 Alcalá de Henares,
Madrid, Spain.
Fax: 34 91 8854763
francisco.mendicuti@uah.es

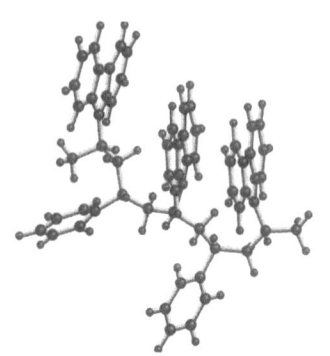

Specialty Keywords: Excimers, Energy transfer, Polymers, Inclusion Complexes, Molecular Mechanics, Molecular Dynamics.

We apply the steady state and time-resolved fluorescence techniques, as well as various theoretical methods for the study of some conformational properties in polymer systems and the inclusion processes of small molecules and polymers with cyclodextrins. Comparison of the theoretical and experimental results allow us to learn more about the conformations and dynamics of polymeric systems and the driving forces and thermodynamics accompanying complexation processes.

Gallego, J., Pérez-Foullerat, D., Mendicuti, F., Mattice, W.L. *J. Polym. Sci. Polym.Phys. Ed* **2001**, 39, 1272.
Pastor, I.,Dimarino, A., Mendicuti, F. *J. Phys. Chem. B* **2002**, 106(8), 1995.

Date submitted: 3rd September 2002   **Fabienne Mérola, Ph.D.**

Laboratory of Chemical Physics, Bat 349,
Université Paris-Sud,
Orsay, F-91405,
France.
Tel: +33 1 69 15 30 17   Fax: +33 1 69 15 61 88
fabienne.merola@lcp.u-psud.fr
www.lcp.u-psud.fr

Specialty Keywords: Protein dynamics, Cell signaling, Time-resolved spectroscopy.

I work to the development of new cell imaging and diagnosis methods based on time-resolved fluorescence: a thorough background in the physics and chemistry of proteins in solution is the basis for original approaches of their dynamics and interactions inside the living cell. We use single and two-photon laser excitation, combined with FRET and FLIM techniques, to investigate the regulation of ion channels involved in synaptic communication and muscle contraction, and, more recently, the structure-photophysics relationship in fluorescent proteins.

Martinez et al. (2002) "Allosteric transitions of *Torpedo* acetylcholine receptor in lipids, detergent and amphipols: molecular interactions vs. physical constraints", *FEBS Lett.* in press
Guiot et al. (2000) "Molecular dynamics of biological probes by fluorescence correlation microscopy with two-photon excitation", *J. Fluorescence* **10**, 413-419.

Date submitted: 30th August 2002

Svetlana B. Meshkova, D.Sc., Ph.D.

Department of Analytical Chemistry and Physico-Chemistry of Coordination Compounds, A.V. Bogatsky Physico-Chemical Institute of National Academy of Sciences of Ukraine, National Academy of Sciences of Ukraine.
86, Lustdorfskaya doroga, 65080, Odessa, Ukraine.
Tel: +38(0482) 652 042   Fax: +38(0482) 652 012
physchem@paco.net

Specialty Keywords: Fluorescence, Energy Transfer, Lanthanide Complexesl.

Current Research Interests: Design and investigation of photochemical properties of lanthanide complexes in solution and solid matrix. Investigation of connection between the composition, stability and optical properties of complexes and characteristics of lanthanide ions and ligands. Study of new means for elimination of intra- and intermolecular energy losses and its realization in luminescent analysis.

S.B. Meshkova (2000). The Dependence of the Luminescence intensity of Lanthanide Complexes with β-Diketones on the Ligand Form: J. of Fluorescence, 10(4), 333-337.

S.B. Meshkova, Z.M. Topilova, D.V. Bolshoy, S.V. Beltyukova, M.P. Tsvirko and V.Ya. Venchikov (1999). Quantum Efficiency of the Luminescence of Ytterbium (III) β-Diketonates: Acta Phys. Polonica A, 95(6), 983-990.

Date submitted: 20th June 2002

Olaf Minet, Ph.D.

Institute for Medical Physics,
Freie Universität Berlin,
Fabeckstr. 60-62, 14195 Berlin, Germany.
Tel: +49 30 84492311   Fax: +49 30 84494377
minet@zedat.fu-berlin.de
www.fu-berlin.de

Specialty Keywords: Optical Biopsy, Optical Molecular Imaging, Image processing and analysis.

Current Research Interests: My research is focused on advancing fluorescence applications in medicine. This involves native autofluorescence compounds like NADH in Optical Biopsy and synthetic markers in Optical Molecular Imaging as well. Of special interest are investigations in the field of image processing, i.e. for eliminating the effects of tissue optics like absorption and scattering on the fluorescence signal, also called rescaling.

J. Beuthan, O. Minet (1999): Fluorescence Diagnosis in the Border Zone of Liver Tumors. In: W. Rettig et al. (eds.) Applied Fluorescence in Chemistry, Biology and Medicine. Springer. Berlin, Heidelberg, N.Y., 537-551.

Date submitted: 12th August 2002

Anatolii G. Mirochnik, Ph.D.

Far-Eastern Branch of the Russian Academy of Sciences,
Institute of Chemistry,
159 prosp.100-letiya Vladivostoka, Vladivostok,
690022, Russia.
Tel: (4232) 310466   Fax: (4232) 311 889
mirochnik@ich.dvo.ru

Specialty Keywords: Fluorescence, Polymer photochemistry.

Design and investigation of fluorescence and photochemical properties of lanthanide and p-elements (boron, s^2 – ions) complexes. Study of photochemical reaction mechanisms, ascertainment of correlations between spectroscopic parameters and molecular structure.

Mirochnik A.G., Gukhman E.V., Zhihareva P.A., Karasev V.E. Excimer Formation of Dibenzoylmethanatoboron Difluoride During Photolysis in Polymer Films 2002, Spectroscopy Letters, **35,** 309-315.
Mirochnik A.G., Petrochenkova N.V., Karasev V.E. Enhancement of Luminescence in the Photolysis of Eu(III) poly(acrylic acid-co-butyl methacrylate) complexes1998, Spectroscopy Letters, **31,** 1167-1177.

Date submitted: 13th September 2002

Hirdyesh Mishra, Ph.D.

Photophysics laboratory,
Department of Physics,
Kumaun University,
Nainital – 263 002, India.
Tel: +91 5942 37450(O), 32757(R)   Fax: +91 5942 35576
hirdyesh@yahoo.com

Specialty Keywords: Time-domain fluorescence spectroscopy of H-bonded molecular system and its applications, Theoretical computation, Instrumentation.

Research Interest: My basic research interest is to understand various photo-induced electronically excited state relaxation processes viz. ESPT, ET, TICT, EERS etc through experimental and theoretical investigations and its applications as fluorescence sensors, lasing materials, luminescence collectors, memory devises etc. in some hydrogen bonded molecular system in polymers. Besides this I am also interested to design and fabrication of instruments and programming for computation.

An optical approach for sensing pH based on energy transfer in nafion matrix. V. Mishra, H. C. Joshi and T.C. Pant: Sens. Accut. 82 (2002) 133-141

Date submitted: 12th August 2002

Tom Misteli, Ph.D.

National Cancer Institute, NIH,
41 Library Drive, Bldg. 41, B610,
Bethesda, MD 20892,
USA.
Tel: 301 402 3959   Fax: 301 496 4951
mistelit@mail.nih.gov
rex.nci.nih.gov/RESEARCH/basic/lrbge/cbge.html

Specialty Keywords: Living cells, Photobleaching, Modelling.

My laboratory uses photobleaching, in situ hybridization and FCS methods to study nuclear architecture and genome expression in vivo. We make heavy use of kinetic modeling methods to analyze in vivo microscopy data.

Phair R.B and T. Misteli, High mobility of proteins in the mammalian cell nucleus. Nature, 404, 604-609 (2000)
Phair R.B. and T. Misteli, Kinetic modeling approaches to in vivo microscopy, Nature Rev. Mol. Cell Biol., 2, 898-907 (2001)

Date submitted: 12th August 2002

Ihab Kamal Mohamed, Ph.D.

Zoology Dept., Faculty of Science,
Ain-Shams Uni., Cairo, Egypt, Tel. +20 (2) 6390470
& Cell biology (Ls. Plattner), biology Dept.,
Konstanz Uni., Germany.
ihabkmohamed@yahoo.com
ihab.Mohamed@uni-konstanz.de
www.ub.uni-konstanz.de/kops/volltexte/2002/760

Specialty Keywords: Ca^{2+}, Exocytosis, Secretion, Fluorochrome analysis, Paramecium.

Current Research: Cellular calcium signaling during secretion. This is proceeded under CLSM or 2λ inverted microscope by painstaking Ca^{2+}-sensing fluorochrome microinjection into individual living cells and stimulation of these cells. Then transform the detected fluorescence change into a calcium quantitative values vs. time (time-resolved fluorescence imaging) by sophisticated computerized process. I am also, interested in new biological fluorescence sensing methodology e.g. fluorochrome microinjection, GFP application, imaging single molecules, signal transduction and looking for a post-doctor position in that field.

B.Sc, M.Sc. Ain-Shams, London, Wales Universities, Ph.D. (2002) Konstanz Uni. Germany.
* I. Mohamed et al. (2002), J. Membrane Biol. 187, 1-14.

Date submitted: 5th August 2002

Gerhard J. Mohr, Ph.D.

Institute of Physical Chemistry,
Friedrich-Schiller University,
Lessingstrasse 10, D-07743 Jena,
Germany.
Tel: +49 3641 948379   Fax: +49 3641 948302
gerhard.mohr@uni-jena.de
www.uni-jena.de/chemie/institute/pc/grummt/mohr_home.htm

Specialty Keywords: Luminescent sensors, Optodes, Reactands, Labels.

Current research is dedicated to the development of new functional dyes and the investigation of their sensing properties in thin polymer layers. Then, they are adapted to miniaturized optics components for the detection of gaseous and dissolved analytes relevant in environmental, medical and biotechnical areas. Furthermore, we develop novel long-wavelength absorbing and fluorescing dyes that can be used for labeling of biomolecules.

G. J. Mohr et al. (1999). Reversible chemical reactions as the basis of optical sensors used to detect amines, alcohols and humidity, *J. Mat. Chem.* **9**, 2259-2265.
P. Czerney, F. Lehmann, M. Wenzel, V. Buschmann, A. Dietrich, G. J. Mohr. (2001). Tailor-made dyes for Fluorescence Correlation Spectroscopy, *Biol. Chem.* **382**, 495-498.

Date submitted: 5th September 2002

María C. Moreno-Bondi, Ph.D.

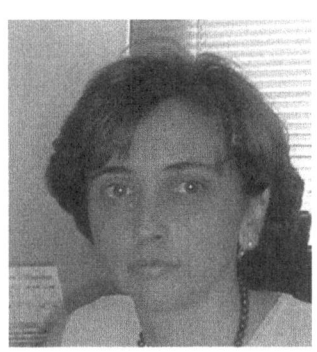

Dept. Analytical Chemistry, Facultad de Química,
Complutense University,
Madrid, E-28040,
Spain.
Tel: +34 913944196   Fax:+34 913944329
mcmbondi@quim.ucm.es
www.ucm.es/info/analitic/

Specialty Keywords: Opt(r)odes, Luminescent sensors, Analysis, Validation.

Our current areas of reseach are (i) the design, fabrication and analytical characterization of *fiber optic* chemosensors *and* biosensors based on novel dyes for the analysis of environmental, industrial, food and medical parameters; (ii) the synthesis and application of Molecularly Imprinted Polymers (MIPs) for sensor development and separation purposes; (iii) sensor application and validation.

M.P. Xavier, B. Vallejo, M.D. Marazuela, M.C. Moreno-Bondi, F. Baldini, A. Falai, *Biosens. and Bioelect.* **2000**, *14*, 895.
M. Bedoya, G. Orellana, M.C. Moreno-Bondi, *Helv. Chim. Acta* **2001**, *84*, 2628.

Date submitted: 13th September 2002

Larry E. Morrison, Ph.D.

Research and Development, Vysis, Inc.,
3100 Woodcreek Drive, Downers Grove,
Illinois, 60515,
USA.
Tel: 630 271 7136   Fax: 630 271 7128
lmorrison@vysis.com

Specialty Keywords: Fluorescence, In situ hybridization, Energy transfer assays.

Current Research Interests: Developing diagnostic, prognostic, and predictive assays for human cancers employing both fluorescence *in situ* hybridization and PCR-based assays. This has included developing multi-target *in situ* hybridization technology using many fluorescent labels simultaneously, combinatorially, or ratiometrically. An early and continuing interest is homogeneous fluorescence detection systems, especially as applied to detecting PCR products.

Morrison (1999) Homogeneous detection of specific DNA sequences by fluorescence quenching and energy transfer, Journal of Fluorescence 9(3) 187-196.

Morrison & Legator (1999) Multi-color fluorescence in situ hybridizations tech. *In* An Intro to Fluorescence *in situ* Hybridization. Andreeff and Pinkel, eds. Wiley-Liss, New York, pp 77-118.

Date submitted: 28th August 2002

Francis Mueller, Ph.D.

F. Hoffmann-La Roche Ltd,
Pharmaceutical Research, Discovery Technologies,
CH-4070 Basel,
Switzerland.
Tel: +41 (0) 61 688 64 30   Fax: +41 (0) 61 688 74 08
francis.mueller@roche.com

Specialty Keywords: Proteins, Binding affinities, Time-resolved Fluorescence labels.

Biomolecular structure research: Protein dynamics, stopped-flow measurements, mobility of tryptophanes for structural studies. Support of fluorescent biological assays development. Intracellular calcium.

Characterisation of lead structures for protein binding. Hits validation from high throughput screening and biological assays. Support in fine tuning of potential ligands with quantitative measurement of affinities by fluorescence titration.

Date submitted: 13th September 2002

Gerhard Müller-Newen, Ph.D.

Institut für Biochemie,
Universitätsklinikum Aachen,
Pauwelsstraße 30,
52057 Aachen, Germany.
Tel: +49 (0)241 80 88860   Fax: +49 (0)241 80 82428
mueller-newen@rwth-aachen.de

Specialty Keywords: Fluorescent fusion proteins, Living cells, Confocal laser-scanning microscopy.

Current research: cytokine signal transduction in live cells using confocal microscopy. To achive this, cytokines, cytokine receptors, Janus kinases and transcription factors of the STAT-family are expressed as fusion proteins linked to GFP, YFP or CFP. The proteins are studied by FLIP (fluorescence loss in photobleaching), FRAP (fluorescence recovery after photobleaching) and FRET to learn more about their subcellular distribution, their dynamics and interactions within the living cell. Since we entered the field of fluorescent proteins just two years ago, the following references refer to former work the group.

Müller-Newen, G., A. Küster, J. Wijdenes, F. Schaper, P. C. Heinrich. 2000. Studies on the IL-6-type cytokine signal transducer gp130 reveal a novel mechanism of receptor activation by monoclonal antibodies. *J. Biol. Chem.* 275: 4579-4586.

Date submitted: 30th August 2002

Kiyofumi Murakami, Ph.D.

Faculty of Education, Yamaguchi University,
Yashida 1677 1,
Yamaguchi 753 8513,
Japan.
Tel: +81 83 933 5351   Fax: +81 83 933 5351
murakami@edu.yamaguchi-u.ac.jp

Specialty Keywords: Biomacromolecule-Small Molecule Interaction, Kinetics and Mechanism.

Current Research Interests: I am interested in specific and cooperative bindings of amphiphilic substances such as dyes and surfactants to biomacromolecules and their local structure models from thermodynamic and kinetic view points. I am also interested in exploring new materials for science education.

K. Murakami (2002). Thermodynamic and kinetic aspects of self-association of dyes in aqueous solution. *Dyes and Pigments*, **53**(1), 31-43. K. Murakami (1999). Cooperative ligand binding to globular protein: A statistical mechanical theory based on a simple geometrical model and its application to lysozyme sysytems. *Langmuir*, **15**(12), 4270-4275.

Date submitted: 12th September 2002   **Miloš Nepraš, Ph.D.**

Department of Organic Technology,
University of Pardubice,
Studentská 95, 532 10 Pardubice,
Czech Republic.
Tel: +420 466038500 Fax: +420 466038004
Milos.Nepras@upce.cz

Specialty Keywords: Fluorescent probes, Bifluorophoric systems, Structure and fluorescence characteristics.
Syntheses and study of relationships between electronic structure and luminescence properties of polynuclear aromatic ketones and quinones and their derivatives. Syntheses and fluorescence characteristics (spectra, quantum yield, fluorescence decay kinetics and solvent effect) of new fluorescent probes derived form acyl and triazinyl derivatives of pyrene, aminopyrenes and aminobenzanthrones. Study of the excitation energy transfer at bifluorophoric systems created form the 3-aminobenzanthrone and aromatic hydrocarbon subsystems.

V. Fidler, P. Kapusta, M. Nepraš, J. Schroeder, I. V. Rubtsov and K. Yoshihara Femtosecond Fluorescence Anisotropy Kinetics as a Signature of Ultrafast Electronic Energy Transfer in Bichromophoric Molecules Z. Phys. Chem. 216 (2002) 589 – 603.

Date submitted: 12th July 2002   **Walter D. Niles, Ph.D.**

Genoptix, Inc., Systems and Applications,
3398 Carmel Mountain Rd., San Diego,
CA, 92037,
USA.
Tel: 858 523 5059 Fax: 858 523 5070
wniles@genoptix.com

Specialty Keywords: Radiometric imaging, Energy transfer, Membrane dynamics.
Developed quantitative fluorescence resonance energy transfer imaging of membrane dynamics in model and biological systems for understanding essential biophysical mechanisms.   Now applying novel fluorescence and optical micromechanics for development of assay technologies (biologies and instrumentation) for drug discovery and diagnostics.

Endothelial cell-surface gp60 activates vesicle formation and trafficking via Gi-coupled Src kinase signaling pathway. 2000. Journal of Cell Biology 150:1057-1069.
Radiometric calibration of a video fluorescence microscope for the quantitative imaging of resonance energy transfer. 1995. Review of Scientific Instruments. 66:3527-3536.

Date submitted: 15th July 2002

Christopher G. Norey, Ph.D.

Amersham Biosciences, The Maynard Centre,
Forest Farm, Whitchurch,
Cardiff, CF14 7YT,
Wales, UK.
Tel: +44 (0) 29 2052 6439   Fax: +44 (0) 29 2052 6230
christopher.norey@uk.amershambiosciences.com
www.amershambiosciences.com

Specialty Keywords: Polarization; Assays; HTS Instrumentation.

Our interests are development of systems relevant for high throughput screening assays, employing fluorescence polarization, FRET and time resolved-FRET techniques. Primarily using CyDye™ fluors and Eu (TMT) chelates with detection via single well PMT readers or whole plate imaging platforms, such as LEADseeker™ multi-modality imaging system. We have a particular interest in receptor ligand interactions, protease cleavage and kinase assays. Recently we have been investigating the application of fluorescence lifetime to these areas.

A. Fowler, I. Davies and C. Norey, (2000), A Multi-Modality Assay Platform for Ultra-High Throughput Screening. *Current Pharmaceutical Biotechnology*, **1**, 265-281.

A. Harris, S. Cox and C. Norey, (2002), High-throughput fluorescence polarization receptor binding assays. In: *LifeScience News*, Amersham Biosciences UK Limited, issue 10, 17-19.

Date submitted: 12th September 2002

Mercedes Novo, Ph.D.

Universidad de Santiago de Compostela,
Facultad de Ciencias, Departamento de Química Física,
Campus Universitario s/n, E-27002 Lugo,
Spain.
Tel: +34 982 223325   Fax: +34 982 22 49 04
mnovo@lugo.usc.es

Specialty Keywords: Fluorescence, Data analysis.

Current interests: Study of the influence of confined media such as cyclodextrins on proton transfer and charge transfer processes. Design of fluorescent probes for the characterisation of supramolecular structures formed by cyclodextrins. Development and implementation of new data analysis methods for steady state and time resolved fluorescence data.

W. Al-Soufi, M. Novo y M. Mosquera (2001). Principal Component Global Analysis of fluorescence and absorption spectra of 2-(2'-hydroxyphenyl)benzimidazole. *Appl. Spectrosc.,* **55**, 630-636. E. Alvarez-Parrilla, W. Al-Soufi, P. Ramos Cabrer, M. Novo y J. Vázquez Tato (2001). Resolution of the association equilibria of 2-(p-toluidinyl)-naphthalene-6-sulfonate (TNS) with cyclodextrin and a charged derivative. *J. Phys. Chem. B*, **105**, 5994-6003.

Date submitted: 26th August 2002

Guillermo Orellana, Ph.D.

Laboratory of Applied Photochemistry, Faculty of Chemistry,
Universidad Complutense Madrid,
28040 Madrid,
Spain.
Tel: +34 913944220 Fax: +34 913944103
orellana@quim.ucm.es
www.ucm.es

Specialty Keywords: Indicator design, Fiber-optic sensors, Environmental analysis and Process control.

Our currents areas of research are (i) design and fabrication of **micro-probes** based on molecularly engineered luminescent dyes, novel photochemical reactions and *fiber-optic chemosensors* for in situ analysis of environmental, industrial, and medical parameters, and (ii) synthesis and characterization of **nano-probes** to investigate the structure of nucleic acids and design artificial photonucleases, The realization of both goals rests on *tailored* luminescent transition metal complexes and organic heterocyclic structures.

F. Navarro-Villoslada, G. Orellana, M.C. Moreno-Bondi, T. Vick, M. Driver, G. Hildebrand and K. Liefeith, *Anal. Chem.* **2001**, *73*, 5150-5156.

M.E. Jiménez, G. Orellana, F. Montero and M.T. Portolés, *Photochem. Photobiol.* **2000**, *72*, 28-34.

Date submitted: 24th June 2002

Uwe Ortmann

PicoQuant GmbH,
Rudower Chaussee 29,
Berlin 12489,
Germany .
Tel: +49 (0) 30 6392 6560 Fax: +49 (0) 30 6392 6561
ortmann@pq.fta-berlin.de
www.picoquant.com

Specialty Keywords: Pulsed Laser Systems, Photon Counting Equipment, Time-resolved Fluorescence Systems, Single Molecule Detection Microscopes.

Current Status: Head of Systems and Sales / Marketing divisions of PicoQuant GmbH.

My major activities are based on the design of fluorescence system, especially in the field of time-resolved spectrometer and single molecule detection.

Date submitted: 9th September 2002

Martin H. Otz, Ph.D.

Syracuse University, Dept. of Earth Sciences,
313 Heroy Geology Laboratory, Syracuse,
Onondaga, 13244-1070,
USA.
Tel: 315 572 0254
otzhydro@hotmail.com
web.syr.edu/~mhotz/index.html

Specialty Keywords: Dye tracing, Hydrogeology, Fluorescent dyes.

A major problem in hydrology is to determine the flow paths of water in organic-rich environments. My research focuses on the development of dye tracing techniques for tracing and quantifying mixtures of organic-rich waters using organic fluorescent dyes.

Otz, M.H., Otz, H.K., and Keller, P., 2002, Detection limits for spectro-fluorometry: a case study in the region of Finstersee (ZG), northern Switzerland [abs.]: EOS (Transactions American Geophysical Union), v.83. p. S-183.

Otz, M.H., Hanselmann, K., Otz, H.K., Tonolla, M., and Siegel, D.I., 2000, Is the biocline of meromictic Lake Cadagno (Swiss Alps) affected by complex lake current patterns? [abs.]: Eos (Transactions American Geophysical Union), v. 81, F-473.

Date submitted: 9th August 2002

Roger H. Pak, Ph.D.

Macromolecular Structure Dept,
Bristol-Myers Squibb Pharmaceutical Research Institute,
5 Research Parkway, Wallingford, CT,
06492, USA.
Tel: 203 677 7225   Fax: 203 677 6984
roger.pak@bms.com

Specialty Keywords: Bioconjugate / Biophysical Chemistry, Biomolecular Assay Design and High-Throughput Screening.

My research focuses on developing labeled peptides and bioconjugates for use in biomolecular assays and high-throughput screening for drug discovery. These bioconjugates are used in a variety of assay formats such as time-resolved fluorescence resonance energy transfer, fluorescence polarization, fluorescence intensity and other radioisotopic or luminescent techniques such as scintillation proximity assays, bioluminescence and enzyme-coupled reactions. I am also involved in the development of novel fluorophores as biological and chemical sensors.

Pantano, P.
Papageorgiou, G.C.

Date submitted: 14th August 2002

PantanoLABO
est. 1996

Paul Pantano, Ph.D.

Department of Chemistry,
The University of Texas at Dallas,
Richardson, TX 75083-0688,
USA.
Tel: 972 883 6226 Fax: 972 883 2925
pantano@utdallas.edu
www.utdallas.edu/dept/chemistry/faculty/pantano.html

Specialty Keywords: Microarrays, Sensors, Cell adhesion.
PantanoLABO is motivated to develop elegant analytical techniques and methodologies to understand complex (bio)chemical systems. Our research includes the fabrication and characterization of microwell, micropost, nanotip, and planar imaging fiber chemical and electrochemical sensors. Specific biological interests include cell adhesion and guidance, reactive oxygen species and oxidative stress, and neurochemical dynamics. New projects include immunosensor arrays, cell-based biosensors, and other high-throughput screening assays.
C. C. Meek and P. Pantano, (2001). Spatial Confinement of Avidin Domains in Microwell Arrays, *Lab on a Chip*, **1** (2), 158-163.
E. S. Jin, B. J. Norris, and P. Pantano, (2001). An Electrogenerated Chemiluminescence Imaging Fiber Electrode Chemical Sensor for NADH, *Electroanalysis*, **13** (15), 1287-1290.

Date submitted: 3rd September 2002

George C. Papageorgiou, Ph.D.

National Center for Scientific Research Demokritos,
Institute of Biology,
Athens,
Greece. 153 10.
Tel: 3010 650 3551 Fax: 3010 651 1767
gcpap@bio.demokritos.gr & gcpap@ath.forthnet.gr

Specialty Keywords: Photosynthesis, Chlorophyll, Cyanobacteria.

Recently, we have explored applications of phycobilisome-sensitized chlorophyll *a* fluorescence as a quantitative reporter of osmotic volume changes of cyanobacteria, and of osmotically-driven transport of solutes and water across cyanobacterial cell envelopes.

Stamatakis K and Papageorgiou GC (2001) The osmolality of cell suspension regulates phycobilisome-to-photosystem I excitation transfer in cyanobacteria. Bioch. Biophys. Acta 1506: 172-181.
Stamatakis K, Ladas Np, Alygizaki-Zorba A and Papageorgiou GC (1999) Sodium chloride-induced volume changes of freshwater cyanobacterium Synechococcus sp PCC7942 cells can be probed by chlorophyll a fluorescence. Arch. Biochem. Biophys. 370: 240-249.

Date submitted: 9th September 2002　　**Vladislav Papper, Ph.D.**

Institute of Chemistry, Humboldt University of Berlin,
Brook-Taylor Strasse 2,
Berlin, 12479,
Germany.
Tel: +49 30 20937133
vladp@rz.hu-berlin.de
www.chemie.hu-berlin.de/wr/index.html

Specialty Keywords: Stilbene, Photoisomerisation, Dual Fluorescence.

Synthesis, photochemistry and photophysics of stilbenoid compounds, mainly *trans-cis* photo-isomerisation. Synthesis and photophysics of fluorescent and dual-fluorescent probes, derivatives of stilbene, with applications to biological membranes, proteins of biological interest, polarity probes for quarz surfaces with the following application to the optoelectronic devices. Synthesis, photophysics and photochemistry of dual-fluorescent probes, *p-(N,N-dimethylamino)benzonitrile* derivatives, for visual and proton-pumping opsin proteins.

V. Papper, G. I. Likhtenshtein, *"Substituted Stilbenes: A New View on Well-Known Systems"*, J. Photochem. Photobiol. A: Chem. 140, (2001), 39-52.

V. Papper, V. Kharlanov, W. Rettig, *"New fluorescent probes for visual proteins"*, Phys. Chem. Chem. Phys. 4, (2002), 1752 – 1759.

Date submitted: 18th August 2002　　**Alexandr S. Parfenov, Ph.D.**

Department of Biochemistry and Molecular Biology,
Center for Fluorescence Spectroscopy University of Maryland,
725 West Lombard Street, Baltimore,
21201, USA.
Tel: 410 706 8409 Fax: 410 706 8408
alexandrparfenov@yahoo.com

Specialty Keywords: Non-invasive diagnostics, Glucose, Cholesterol.

Method of determining skin tissue cholesterol US Patent 6,365,363 Apr.2, 2002.

Fluorescence method for monitoring of glucose in interstitial fluids. SPIE 2001, 4263.
To continue working as a scientist in the field of the non-invasive diagnostics on the development of new diagnostic tests.

Parkhomyuk–Ben Arye, P.
Parola, A.H.

Date submitted: 28th August 2002

Pavel Parkhomyuk–Ben Arye, M.Sc.

Ben-Gurion University,
Department of Chemistry,
Beer-Sheva, 84152,
Israel.
Tel: 972 8 6472189
parhomyu@bgumail.bgu.ac.il
www.bgu.ac.il/chem/index.html

Specialty Keywords: Fluorescence-Based Sensors, FRET, Biophysical Chemistry.

Current Research Interests: (a) Application of FRET for quantitative analysis at nanomolar scale, (b) investigation of the surface phenomena with covalently immobilized fluorescent probes, (c) study of the double spin-fluorescent molecules and their application as redox and viscosity probes and (d) photophysical and photochemical investigation of HSA-Hemin complex.

P. Parkhomyuk-Ben Arye, N. Strashnikova, G.I. Likhtenshtein (2002). Stilbene photochrome-fluorescence-spin molecules: covalent immobilization on silica plate and applications as redox and viscosity probes, *J. Biochem. Biophys. Methods*, **51**, 1-15.

Date submitted: 4th September 2002

Abraham H. Parola, Ph.D.

Chemistry, Ben-Gurion University,
P.O. Box 653, Beer Sheva,
Israel, 84105.
Tel: 972 8 6472454   Fax: 972 8 6472943
aparola@bgumail.bgu.ac.il

Specialty Keywords: Lipid-Protein & Protein-Protein & Protein-Ligand / drug Interactions, Membrane Dynamics, Time / phase resolved fluorescence spectroscopy.

Research topics: the role of hydrophobic interactions in membranal and non-membranal protein function and regulation, signal transduction, cell cycle and proliferation, cell differentiation and intercellular interactions, angiogenesis, apoptosis, magnetic field effects on biological systems.
On the Regulatory Role of Dipeptidyl Peptidase IV (= CD26 = Adenosine Deaminase Complexing Protein) on Adenosine Deaminase activity. I. Ben-Shooshan, A. Kessel, N. Ben-Tal, R. Cohen-Luria and A.H. Parola .*Biochim. Biophys. Acta*, 1587, 21-30 (2002).
Nature of interaction between basic fibroblast growth factor and the antiangiogenic drug 7,7-(carbonyl-bis[imino-N-methyl-4,2-pyrrolecarbonylimino[N-methyl-4,2-pyrrole]-carbonylimino])-bis-(1,3-naphtalene disulfonate).
Removal of polar interactions affects protein folding. M. Zamai, C. Hariharan, D. Pines, M. Safran, A. Yayon, V.R. Caiolfa, R. Cohen-Luria, E. Pines and A.H. Parola. *Biophys. J.*, in press.

Date submitted: 6th September 2002

Jana Peknicova, Ph.D.

Dept. of Biology and Biochemistry of Fertilization,
Institute of Molecular Genetics Academy of Scences of the
Czech Republic, Videnska 1083, Prague 4,
142 20, Czech Republic.
Tel: + 420 2 44471707
jpeknic@biomed.cas.cz
www.img.cas.cz

Specialty Keywords: Biology of Reproduction, Fertilization, Sperm proteins.

The long-term interest of the group lies in studies of the molecular mechanism of mammalian fertilization. The role of selected sperm proteins during capacitation, acrosome reaction and sperm binding to zona pellucida of oocytes is studied. The changes in immunochemical localization of cytoskeletal proteins in boar sperm during capacitation and acrosome reaction were tested. The effect of endocrine disruptors on mammalian fertility was also tested and sperm quality was evaluated with monoclonal antibodies by immunofluorescence method.

Peknicova J., Kubatova A., Sulimenko V., Draberova E., Viklicky V., Hozak P., Draber P.: Biology of Reproduction 65:672-679, 2001 .
Peknicova J., Kyselova V., Buckiova D., Boubelik M.: American Journal of Reproductive Immunology 47: 311-318, 2002.

Date Submitted: 24th May 2002

Fabrizio Pelella, (Ph.D. Student)

Institute of Protein Biochemistry,
Via Pietro Castellino, 111
Naples, 80131,
Italy.
Tel: +39 0816132312   Fax: +39 0816132270
pelella@dafne.ibpe.na.cnr.it

Specialty Keywords: Biosensors, Thermophilic Proteins and Enzymes, Fluorescence.

My scientific interests deal with the development of innovative protein biosensors based on the utilization of proteins and enzymes isolated from mesophilic and thermophilic organisms. My primary goal is to contribute to the realization of new fluorescence methods of sensing by means of fluorescence techniques. In particular my thesis is focused on the development of stable and non-consuming substrate biosensors for analytes of high environmental, clinical and social interests.

Date submitted: 8th September 2002

Philippe Peltie, Ph.D.

DIRECTION DE LA
RECHERCHE
TECHNOLOGIQUE

CEA / GRENOBLE,
Dept. Systems Pour L'Information et la Sante,
17 rue des martyrs,
38054 Grenoble cedex 9.
Tel: 33 4 38 78 38 12   Fax: 33 4 38 78 57 87
philippe.peltie@cea.fr

Specialty Keywords: Fluorescence instrumentation, Quantitative fluorescence microscopy, DNA.

My job, at present time, is concerned about fluorescence instrumentation, in two domains: Quantitative fluorescence microscopy for silicon DNA biochips (as MICAM™ chip) reader And fluorescence polarization in microchannel's labs on chip for genotyping. In the future, two ways are explored: fluorescence microscopy for cells on chip and in vivo Fluorescence of human tissues through endoscopic way.

Fluorescence detection for DNA chips and labs on chips and perspective for integrated systems. IXth international symposium on luminescence spectrometry in biomedical and environmental Analysis; may, 15-17, 2002; Montpellier, France.

Date submitted: 30th May 2002

Michael J. Pender, M.S.

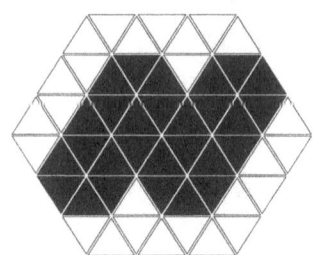

Nanochron, LLC
4201 Wilson Blvd. #110-615,
Arlington, Virginia 22203, USA.

Michael.Pender@Nanochron.com
www.nanochron.com

Specialty Keywords: Photonics, Predictive modeling.

My work focuses on the development of application-specific optical devices. Specific topics include intra-molecular photonic transfer in fluorescent and quasi-fluorescent optical channels and predictive modeling of the properties of fluorophores in photonic devices for optical communications and signal processing.

M. Pender (2001). Optical matrix photonic logic device and method for producing the same, Patent Cooperation Treaty Application No. PCT/IB01/00888.

Date submitted: 30th August 2002   **Xinzhan Peng, Ph.D.**

LI-COR BioSciences,
Division of Chem. R&D,
4308 Progressive Ave, Lincoln,
Nebraska, 68504, USA.
Tel: 402 467 0796   Fax: 402 467 0819
xpeng@licor.com
www.licor.com

Specialty Keywords: Fluorescent probe, Near-infrared dye, Protein assay.

My current research focuses on the design and synthesis of new fluorescent dyes for bio-molecules conjugation. Particular interest is the design and development of novel near-infrared fluorescent probes with high sensitivity for protein assay applications.

Date submitted: 31st August 2002   **Alfons Penzkofer, Ph.D.**

Naturwissenschaftliche Fakultät II –Physik,
Universität Regensbur, Universitätsstrasse 31,
Regensburg, D-93053,
Germany.
Tel: +49 941 9432107   Fax: +49 941 9432754
alfons.penzkofer@physik.uni-regensburg.de
www.physik.uni-regensburg.de/forschung/penzkofer

Specialty Keywords: Absorption Spectroscopy, Fluorescence Spectroscopy, Femtosecond Laser Spectroscopy.

We determine refractive index spectra, absorption cross-section spectra, make fluorescence spectroscopic characterisations (fluorescence quantum yields, fluorescence quantum distributions, fluorescence lifetimes, degrees of fluorescence polarization) and study photo-degradation mainly on organic molecules, luminescent polymers, and sensory biological photo-receptors (flavins, bacteriochlorophylls). We perform laser studies on thin-film luminescent polymers and solid-state dye lasers.

W. Holzer, M. Pichlmaier, E. Drotleff, A. Penzkofer, D. D. C. Bradley, and W. J. Blau, Optical Constants Measurement of Luminescent Polymer Films, Opt. Commun., 163 (1999) 24-32.

W. Holzer, M. Pichlmaier, A. Penzkofer, D. D. C. Bradley, and W. J. Blau, Fluorescence Spectroscopic Behaviour of Neat and Blended Conjugated Polymer Thin Films, Chem. Phys. 246 (1999) 445-462.

Date submitted: 22nd August 2002

Frederick S. Perry.

Boston Electronics Corporation,
91 Boylston Street, Brookline,
MA, 02445,
USA.
Tel: (800) 347 5445 or (617566 3821)   Fax: (617) 731 0935
fsp@boselec.com
www.boselec.com

Specialty Keywords: TCSPC, Photodetection, Spectroscopy.

President and founder of Boston Electronics Corporation, North American agents for Becker & Hickl GmbH of Berlin, Germany and for Edinburgh Instruments Ltd of Edinburgh, Scotland. Specialists in photodetection and signal processing electronics for photodetection.

Date Submitted: 13th May 2002

Basilio Pispisa, Ph.D.

Dept.of Chemical Sciences and Technologies,
University of Roma Tor Vergata,
Via Ricerca Scientifica, Rome,
00133 Italy.
Tel: +39 067259 4467   Fax: +39 062020 420
pispisa@stc.uniroma2.it
www.stc.uniroma2.it/files/Pispisa%20files/B.Pispisa

Speciality keywords: Biophysical chemistry, Spectroscopy Conformational analysis.

Three major research topics are pursued in the Professor Pispisa's laboratory:

- structure and molecular dynamics of oligopeptides and polypeptides in solution, mimicking proteins and bioactive compounds;
- structure-reactivity relationships in model compounds of enzymic materials;
- structural features of glycopeptides and functionalized peptides in solution and in membranes.

B. Pispisa et al. (2000) *Biopolymers*, **54**, 127-136. 2002 Peptide-Sandwiched Protoporphyrin Compounds Mimicking Hemoprotein Structures in Solution
B. Pispisa et al. (2002) *J. Phys. Chem.* B 106, 5733-5738. Effects of Helical Distortions on the optical Properties of Amide NH Infrared Absorption in Short Peptides in Solution.

Date submitted: 11th September 2002   **Emmanuelle Plantin-Carrenard, Ph.D.**

Laboratoire de Biochimie Générale et de Glycobiologie,
UFR des Sciences Pharmaceutiques et Biologiques,
Uni. René Descartes - Paris 5, 4 avenue de L'Observatoire,
75006 Paris - France.
Tel: 33 1 53739652   Fax: 33 1 53739655
eplantin@wanadoo.fr

Specialty Keywords: Fluorescence probes, Oxidative stress, Apoptose.

Oxidative stress is defined as the pathological outcome of overproduction of oxidative species that overwhelms the cellular antioxidant capacity. The consequence of induced-oxidative stress on cellular interactions are studied *in vitro* on adherent and non-adherent cell models. Fluorescent probes are interesting tools to measure with high sensitivity and specificity the modifications of cellular mechanisms under oxidant conditions such as capacity of reactive oxygen species production, thiol depletion and apoptosis.

Plantin-Carrenard E. et al. Journal of Fluorescence, 2000 ; 10 : 167-73.
Plantin-Carrenard E. et al. Act Pharm Biol Clin, 2000 ; 11 : 168-73.

Date submitted: 12th September 2002   **Jaromír Plášek, Ph.D.**

Faculty of Mathematics and Physics – Institute of Physics,
Charles University, Ke Karlovu 5,
Prague, CZ-12116,
Czech Republic.
Tel: +420 2 21911349
plasek@karlov.mff.cuni.cz

Specialty Keywords: Membrane potential, Polarized fluorescence, Microfluorimetry.

Research Interests: Lipid order in cell membranes from polarized fluorescence of membrane probes. Fluorescent probing of cell membrane and mitochondrial membrane potential in living cells. ATP binging to a N-domain in the cytoplasmic loop of Na,K-ATPase from binding assays with TNP-ATP.

J. Plášek and K. Sigler (1996) Slow fluorescent indicators of membrane potential: a survey of different approaches to probe response analysis. *J. Photochem. Photobiol. B: Biology* **33**, 101-124.

D. Gášková, R. Čadek, R. Chaloupka, J. Plášek and K. Sigler (2001) Factors underlying membrane potential-dependent and -independent fluorescence responses of potentiometric dyes in stressed cells: diS-C$_3$(3) in yeast. *Biochim. Biophys. Acta* **1511**, 74-79.

Prieto, M.
Procházka, K.

Date submitted: 31st August 2002

Manuel Prieto, Ph.D.

Centro de Química-Física Molecular, IST,
Av. Rovisco Pais,
1049-001 Lisbon,
Portugal.
Tel: +351 21 8419219   Fax: +351 21 8464457
prieto@alfa.ist.utl.pt

Specialty Keywords: FRET, Lipid domains, Lipid protein-
interaction.

Current Research Interests: Application of steady-state and time-resolved photophysical methodologies to the detection, characterization and dynamics of membrane heterogeneities (domains/rafts). Topology and dynamics of protein/peptide and polyene antibiotics interaction with model systems of membranes. Cholesterol organization in membranes.

R. Almeida, L. Loura, A. Fedorov, M. Prieto (2002). Non-equilibrium phenomena in the phase separation of a two-component lipid bilayer. *Biophys. J.,* **82** (2), 823-834 .
L. Loura, R. Almeida, M. Prieto (2001). Detection and characterization of membrane microheterogeneity by resonance energy transfer (review), *J. Fluorescence,* **11**(3), 197-209.

Date submitted: 3rd September 2002

Karel Procházka, Ph.D., D.Sc.

Dept of Physical and Macromolecular Chemistry  Labotory
of Specialty Polymers,
School of Science, Charles University in Prague,
Albertov 6, 128 43 Prague 2, Czech Republic.
Tel: +420 2 21952154   Fax: +420 2 24919752
prochaz@vivien.natur.cuni.cz
www.natur.cuni.cz/pmc

Specialty Keywords: Fluorescence from polymers, Polymer
conformations and Segmental dynamics, Association of
polymers.

Studies of polymer conformations and chain dynamics by a combination of fluorescence techniques (time-resolved and steady-state, recently also FCS) with static and quasielastic light scattering and other techniques used in polymer science (such as SEC, ultracentrifugation, electromigration, SEM and AFM microscopy). In recent decade, a special attention has been paid to the association of amphiphilic water-soluble block copolymers (mainly block polyelectrolytes) in polar and aqueous media.

C. Tsitsilianis, D. Voulgaris, M. Štěpánek, K. Podhájecká, K. Procházka, Z. Tuzar, W. Brown (2000) Polystyrene/Poly(2-vinylpyridine) Heteroarm Star Copolymer Micelles in Aqueous Media and Onion Type Micelles Stabilized by Diblock Copolymers *Langmuir* 16, 6868-6876.

Date submitted: 29th July 2002   **M. Elisabete, C.D. Real Oliveira, Ph.D.**

Physics Department, University of Minho,
Campus de Gualtar, Braga,
Portugal, 4710-057,
Portugal.
Tel: +351 253 604325   Fax: +351 253 678981
beta@fisica.uminho.pt

Specialty Keywords: Biophysics, Microheterogeneous systems binding to DNA.

In the last years my research has been focused to investigate the structural and functional characterization of some nonionic microemulsions and studying nonionic surfactants/lipid interactions using steady state and time resolved fluorescence spectroscopy and fluorescence anisotropy using the fluorescence probes, pyrene, nile red, prodan, and di-asp, DCM laser dye, etc. I am still studying the binding of a cationic pyrene derivative (PMTP) to polynucleotides and DNA and try to investigate the mechanism of the fluorescence quenching involved in the binding process.

"Effect of Surfactants in Soybean Lecithin Liposomes Studied by energy transfer between NBD-PE and N-Rh-PE", A.L.F.Baptista, P.J.G. Coutinho, M.E.C.D. Real Oliveira, J.I.N. Rocha Gomes", *Journal of Liposome Research, 10(4) (2000) 509-519.*

Date submitted: 17th June 2002   **Renata Reisfeld, Ph.D., DHC.**

Department of Inorganic Chemistry,
The Hebrew University,
Jerusalem 91904, Israel.
Tel: 972 2 6585323   Fax: 972 2 6585319
renata @vms.huji.ac.il
chem.ch.huji.ac.il/employee/reisfel/REISFELD.HTM

Specialty Keywords: Fluoresent dyes, Sol-gel tunable lasers, LSC, Sensors, QD.

The group of Prof. Reisfeld is studying the following topics connected with fluorescence. Fluorescence of Rare Earth ions in glasses, theoretically and experimentally, Steady State and dynamic processes of Fluorescence of dyes in glasses. Exited State Process and applications in luminescent solar concentrators (LSC), tunable lasers, planar active wave guides and sensors. Quantum dots (QD)of semiconductors and metals in glass bulks and films. Using absorption and fluorescence, quantum size effects are determined. Applications for nonlinear optics.

R. Reisfeld, "Lasers Based in Sol-Gel Technology", Optical and Electronic Phenomena in Sol-Gel Glasses and Modern Applications, Eds. R. Reisfeld, C.K. Jorgensen, *Structure and Bonding* **85**, Springer- Verlag (1996) 215-233.

Date submitted: 11th July 2002

Ute Resch-Genger, Ph.D.

Project Group I.3902,
Bundesanstalt für Materialforschung und –prüfung (BAM),
(Federal Institute for Materials Research and Testing),
Richard-Willstätter-Str. 11, D-12489 Berlin,
Germany.
Tel: +4930 8104 1134   Fax: +49308104 5005
ute.resch@bam.de
www.bam.de

Specialty Keywords: Fluorescent standards, Fluorescent probes and sensors, Time resolved fluorometry, Quality assurance.

Current Research Interests: Design and spectroscopic study of functional dyes and fluorescent sensor molecules. Quality assurance and standardization including development of fluorescent standards for steady state and time resolved fluorometry.

K. Rurack, U. Resch-Genger (2002). Rigidization, preorientation and electronic decoupling – the magic triangle for the design of highly efficient sensors and switches, *Chem. Soc. Rev.* **31**, 116-127.

Date submitted: 9th September 2002

Wolfgang Rettig, Ph.D.

Institut für Chemie der Humboldt-Universität zu Berlin,
Brook-Taylor-str. 2,
12489 Berlin,
Germany.
Tel: (+49 30) 2093 5585 / 25552   Fax: (+49 30) 2093 5574
rettig@chemie.hu-berlin.de
www.chemie.hu-berlin.de/wr/index.html

Specialty Keywords: Time-resolved fluorescence, Adiabatic photoreactions, TICT.

Mechanisms of photochemical primary processes (electron and proton transfer; trans-cis and valence isomerizations; visual process); ultrafast fluorescence and absorption spectroscopy; solvation of excited states; quantum-chemical modelling of photoreactions; fluorescence probes for biology, medicine and analytical chemistry; fluorescence polymer probing. Many studies enriching the field of compounds with anomalous fluorescence properties linked with intramolecular twisting (TICT).

Applied Fluorescence in Chemistry, Biology, and Medicine", Editors: W. Rettig, B. Strehmel, S. Schrader, H. Seifert, Springer-Verlag Berlin, Heidelberg, 1998.

Date submitted: 12th September 2002

David E. Roll, Ph.D.

Dept. of Chemistry Roberts Wesleyan College,
2301 Westside Drive, Rochester, NY 14624,
Monroe County,
USA.
Tel: 585 594 6485
rolld@roberts.edu
www.roberts.edu

Specialty Keywords: Chlamydia infection, Topoisomerase, Gold nanoparticles.

Type I topoisomerase is an enzyme that plays a role in the regulation of DNA supercoiling in the cell. Research in this lab is directed at understanding the role of this enzyme in the initiation of Chlamydia infection in an eukaryotic cell and the role that phosphorylation may play in regulating this enzyme's activity. In addition, gold nanoparticles and metal enhanced fluorescence may provide valuable tools for the detection of Chlamydia infections.

Date submitted: 6th September 2002

Alexander D. Roshal, Ph.D.

Institute of Chemistry,
at N.V.Karazin Kharkov National University,
4 pl.Svoboda, Kharkov,
Ukraine, 61077.
Tel: (38) 057 245 73 35   Fax: (38) 057 245 71 30
Alexandre.D.Rochal@univer.kharkov.ua
www-chemistry.univer.kharkov.ua/dx/rochal

Specialty Keywords: Flavonoids, Flavonoid complexes, Absorption and Fluorescence spectroscopy.

Research interests: • Structure and physico-chemical properties of flavones, isoflavones and its derivatives. • Proton transfer in flavonols under excitation. • Complexation of flavonoids in the ground and excited states. • Spectral properties of flavonol complexes. • Structure, spectral properties and analysis of pyrylium, benzopyrylium and flavylium salts. • Natural and modified flavonols, coumarins and relative substances as the fluorescent probes for biochemistry and biophysics.

A.D. Roshal, A.V. Grigorovich, A.O. Dorochenko, V.G. Pivovarenko, A.P.Demchenko. *Journal of Physical Chemistry. A.*, **102** (1998), 5907-5914.

Date submitted: 20th July 2002

Victoria V. Roshchina, Ph.D.

Laboratory of Microspectral Analysis of Cells and Cellular Systems,
Russian Academy of Sciences Institute of Cell Biophysics
Avenue Nauki, 3, Pushchino, Moscow Region, 142290
Russia.
Tel: (095) 923 74 67, add.293   Fax: 7 (0967) 79 05 09
lyudam@icb.psn.ru

Specialty Keywords: Plant Physiology and Biochemistry, Sensory Systems, Spectral analysis.

Autofluorescence of intact secretory plant cells is studied in our laboratory. It has been shown that this phenomenon could be recommended: 1) for the diagnostics of secretory cells among non-secretory ones; 2) for the express-analysis of the content of secretory cells at norma and under various factors; 3) for the analysis of cell-cell interactions.

Roshchina V.V., Melnikova E.V. Spectral analysis of intact secretory cells and excretions ofplants. Allelopathy J. 2 (2): 179-188. 1995.

Roshchina V., Melnikova E.V. Microspectrofluorimetry of intact secreting cells, with applications to the study of allelopathy. In: Principles and Practices in Plant Ecology. Allelochemical Interactions. pp.99-126 Ed. Inderjit, Dakshini K.M.M., Foy C.L..Boca Raton, USA :CRC Press. 1999.

Date submitted: 26th August 2002

Anatoly N. Rubinov, Ph.D.

Laboratory of Dye Lasers,
Stepanov Institute of Physics,
70 Skorina Avenue, 220072 Minsk,
Belarus.
Tel: 375 (17) 284 56 24   Fax: 375 (17) 284 16 46
rubinov@ifanbel.bas-net.by

Specialty Keywords: Dye Lasers, Ultrafast Spectroscopy, Fluorescent Probes.

The main results are in the field of ruby and neodymium glass lasers (early 1960s); various types of dye lasers and laser dyes (State Prize of USSR, 1972); laser spectroscopy of organic solutions (State Prize of Belarus, 1994); intracavity laser spectroscopy; distributed-feedback (DFB) lasers including holographic DFB lasers; mode locked dye lasers and time resolved laser spectroscopy of organic molecules in solutions and bio-membranes; interaction of gradient laser fields with biological objects.

Date submitted: 12th September 2002

Angelika Rück, Ph.D.

Institut for Lasertechnologies (ILM),
Helmholtzstrasse 12,
D-89081 Ulm,
Germany.
Tel: +49 (0) 731 1429 16   Fax: +49 (0) 731 1429 42
angelika.rueck@ilm.uni-ulm.de

Specialty Keywords: FLIM, Microspectrofluorometry, PDT.

Development of methods for spectral fluorescence lifetime imaging, based on time correlated single photon counting in combination with laser scanning microscopes for detection and dynamic analysis of signal transduction pathways in living cells during photodynamic therapy (PDT).   Cellular characterization and evaluation of new photosensitizers with one- and two-photon spectral-resolved microscopy. Definition of protein standards for FLIM/FRET measurements of protein interactions in living cells.

A. Rück et al., Light-induced apoptosis involves a defined sequence of cytoplasmic and nuclear calcium release in AlPcS$_4$-photosensitized cells. *Photochem. Photobiol.*, 2000, 72(2): 210-216.
M. Kress and **A. Rück**, Time-resolved microspectrofluorometry and FLIM of photosensitizers using ps pulsed diode lasers in laser scanning microscopes. *J. Biomed. Optics*, accepted.

Date submitted: 22nd August 2002

Knut Rurack, Ph.D.

Dept. I.3902,
Bundesanstalt für Materialforschung und -prüfung (BAM),
(Federal Institute for Materials Research and Testing),
Richard-Willstätter-Strasse 11, D–12489 Berlin,
Germany.
Tel: +4930 8104 5576   Fax: +4930 8104 5005
knut.rurack@bam.de

Specialty Keywords: Fluorescent molecular sensors, Functional dyes, Time-resolved fluorescence.

Current interests: Design of functional dyes, especially fluorescent sensors and switches. Study of charge, electron and proton transfer processes. Investigation of dyes in confined media such as zeolites or sol–gel matrices. Development of fluorescence lifetime standards.

K. Rurack, U. Resch-Genger (2002). Rigidization, preorientation and electronic decoupling—the "magic triangle" for the design of highly efficient fluorescent sensors and switches, *Chem. Soc. Rev.* **31** 116–127.
K. Rurack, A. Koval'chuck, J. L. Bricks, J. L. Slominskii (2001). A simple bifunctional fluoroionophore signaling different metal ions either independently or cooperatively, *J. Am. Chem. Soc.* **123** 6205–6206.

Date submitted: 6th September 2002   **Jean-Marie Ruysschaert, Ph.D.**

Université Libre de Bruxelles,
S.F.M.B. - Structure and Function of Biological Membranes,
Bd. du Triomphe, CP 206/2, Brussels,
1050, Belgium.
Tel: +32 2 650 53 77   Fax: +32 2 650 53 82
jmruyss@ulb.ac.be

Specialty Keywords: Hydrogen/ deuterium exchange,
Fluorescence quenching, Long-range conformational changes.

We have developed a new method to detect changes occuring in the membrane embedded and cytosolic domains of membrane proteins by combining infrared linear dichroic spectra measurements in the course of hydrogen/deuterium exchange with Trp fluorescence quenching by water soluble attenuators. This new approach is of general interest in the study of membrane proteins to detect long-range conformational changes transmitted between the membrane embedded and cytosolic domains.

Grimard V., Vigano C., Margolles A., Wattiez R., van Veen H.W., Konings W.N., Ruysschaert J.-M. and Goormaghtigh E. (2001) Biochemistry 40, 11876-11886.

Date submitted: 20th March 2002   **Alan G. Ryder, Ph.D.**

Department of Physics,
National University of Ireland – Galway,
Galway, Ireland.
Tel: 353 91 750469   Fax: 353 91 750584
alan.ryder@nuigalway.ie
www.physics.nuigalway.ie/People/ARyder/

Specialty Keywords: Time-resolved, Petroleum, Raman.

My research focuses on the use of fluorescence and Raman spectroscopy for the quantitative and qualitative analysis of materials. Specific topics include: using time-resolved fluorescence spectroscopy to analyze petroleum oils, development of lifetime based pH sensors, quantitative measurement of narcotics using Raman spectroscopy, and development of laser diode based time-resolved fluorescence instrumentation.

A.G. Ryder (2002). Quantitative analysis of crude oils by fluorescence lifetime and steady state measurements using 380 nm excitation. *Appl. Spectrosc.* **56**(1), 107-116.

A.G. Ryder (2002). Classification of narcotics in solid mixtures using Principal Component Analysis and Raman spectroscopy., *J. Forensic Sci.* **47**(2), 275-284.

Date submitted: 19th July 2002   **Carlota Saldanha, Ph.D.**

Instituto de Bioquímica, Faculdade de Medicina de Lisboa,
Av. Prof. Egas Moniz, Lisbon,
1649-028 Lisboa,
Portugal.
Tel: +351 217985136   Fax: +351 217939791
saldanha@medscape.com

Specialty Keywords: Acetylcholinesterase, Membrane fluidity, Calcium ion.

Enzyme kinetics studies, namely human erythrocyte and lymphocyte acetylcholinesterase, using fluorescent enzyme substrate and inhibitors. Studies of erythrocyte, lymphocyte and endothelial cells membrane fluidity and erythrocyte exovesiculation using the fluorescent probes diphenylhexatriene, trimethylamino-diphenylhexatriene and hydroxycoumarin. Studies of intracellular second messengers, namely calcium ion and nitrogen monoxide with fluorescent probes.

C. Saldanha and J. Martins-Silva (1996) *Biochem. Educ.* **24**, 235-236.
N. C. Santos, J. Figueira-Coelho, C. Saldanha, and J. Martins-Silva (2002) *Cell Calcium* **31**, 183-188.

Date submitted: 13th September 2002   **Jeffrey S. Sanford, B.S.**

FISH Consultants,
3640 N. Longwood Place,
Tucson, AZ 85750,
USA.
Tel: (520) 241 9656
jsanford@scientist.com
www.FISHconsultants.com

FISH Consultants

Specialty Keywords: FISH, Anatomic Pathology, Automation, Fluorescence Microscopy.

FISH diagnostics and AP laboratory automation HER-2/neu, BCR/abl, PML/RARA, Ploidy Breast, Prostate, and Renal Cancer, Leukemias and Lymphomas.

Wolman SR, **Sanford JS**, Flom K, Feiner H, Abati A, Bedrossian C: Genetic probes in cytology: Principles and Practice, Diagnostic Cytopathology, 13;429-435, 1996.

Micale MA, **Sanford JS**, Powell IJ, Sakr WA, Wolman SR: Defining the Extent and Nature of Cytogenetic Events in Prostatic Adenocarcinoma: Paraffin FISH vs. Metaphase Analysis. Cancer Genetics and Cytogenetics, 69;7-12, 1993.

Date submitted: 19th July 2002

Nuno C. Santos, Ph.D.

Instituto de Bioquímica, Faculdade de Medicina de Lisboa,
Av. Prof. Egas Moniz, Lisbon,
1649-028 Lisboa,
Portugal.
Tel: +351 217985136   Fax: +351 217939791
nsantos@fm.ul.pt

Specialty Keywords: Protein intrinsic fluorescence,
Biomembranes, HIV.

Use of steady-state and time resolved fluorescence spectroscopy (including fluorescence anisotropy, quenching, energy transfer, energy migration and red edge excitation shift) on the study of membrane proteins structure and location, intracellular ion concentration, membrane fluidity, partition of peptides and other fluorescent molecules to biomembranes, erythrocyte membrane vesiculation and binding of small fluorescent molecules to proteins. Characterization of supramolecular systems by light scattering spectroscopy.

N. C. Santos, M. Prieto, and M. A. R. B. Castanho (1998) *Biochemistry* **37**, 8674-8682.
N. C. Santos, J. Figueira-Coelho, C. Saldanha, and J. Martins-Silva (2002) *Cell Calcium* **31**, 183-188.

Date submitted: 4th September 2002

Suzanne F. Scarlata, Ph.D.

Dept. of Physiology & Biophysics,
State University of New York at Stony Brook,
Stony Brook, NY 11794-8661,
U.S.A.
Tel: 631 444 3071   Fax: 631 444 3432

Suzanne.Scarlata@sunysb.edu
physiology.pnb.sunysb.edu/faculty/scarlatta/scarlata.html

Specialty Keywords: Cell Signaling, Protein Assemblies, Protein Dynamics.

Our laboratory has been focusing on the protein-protein interactions that occur in cells during signal transduction. In particular, we have focused on the activation of effector proteins by the subunits of heterotrimeric G proteins. We have studied their in vitro associations on membrane surfaces using fluorescence resonance energy transfer techniques including fluorescence homotransfer. We are currently extending this work to living cells to characterize the localization, protein partners and movement of these proteins upon stimulation.

Date submitted: 16th August 2002

Johannes A. Schmid, Ph.D.

Department of Vascular Biology and Thrombosis Research,
University Vienna,
Brunnerst. 59, A-1235 Vienna,
Austria.
Tel: +43 1 4277 62555   Fax: +43 1 4277 62550
Johannes.Schmid@univie.ac.at
www.univie.ac.at/VascBio/schmid/

Specialty Keywords: FRET, NF-□B, signal transduction.

Current research interests comprise the mechanisms of endothelial cell activation, as well as de-activation, with special focus on the signal-transduction of the NF-B pathway and its interconnection with other pathways. GFP-fusion proteins are used to elucidate the dynamics of signaling molecules in vivo. CFP and YFP-fusion proteins are used to localize protein-interactions in living cells by fluorescence resonance energy transfer microscopy.

J.A. Schmid et al., A. Birbach, R. Hofer-Warbinek, M. Pengg, U. Burner, P.G. Furtmuller, B.R. Binder, R. de Martin R: J. Biol. Chem. 275(22), 17035-42 (2000).
Birbach A., Gold P., Binder B.R., Hofer E., de Martin R., Schmid J.A. J. Biol. Chem. 277(13):10842-51 (2002).

Date submitted: 26th August 2002

Bernhard Schönenberger, Ph.D.

R&D SR, Fluka GmbH,
Industriestrasse 25,
CH-9471 Buchs,
Switzerland.
Tel: 0041 81 755 2609   Fax: 0041 81 755 2736
bschoene@eurnotes.sial.com

Specialty Keywords: Amine-reactive labels, Protein labels, Organic syntheses.

Senior scientist in R&D; in charge of the scientific and technical aspects of the development of the Fluka sales program 'fluorescent markers'. Recent R&D work, partly in cooperation with external groups:
-   chemo- and fluorogenic enzyme substrates for optical sensor applications
-   new fluorescent succinimidylester dyes for multi-labelling of proteins
-   novel assay for succinimidylester purity [1]
-   fluorescence based analytical kits

[1]   Th. Williman, B. Schönenberger, G. Hayenga, Poster P2-076, Euroanalysis 2002, Dortmund, Germany (submitted).

Schwille, P.
Segers-Nolten, I.

Date submitted: 26th August 2002

Petra Schwille, Ph.D.

Experimental Biophysics,
Max-Planck-Institute for Biophysical Chemistry,
Am Fassberg 11, D-37077 Göttingen,
Germany.
Tel: +49 (0) 551 201 1165   Fax: +49 (0) 551 201 1435
pschwil@gwdg.de
www.gwdg.de/~pschwil/

Specialty Keywords: FCS, Two-Photon, Single Molecules.

Development of ultrasensitive fluorescence-based methods for detection and dynamic analysis of single or sparse biomolecules in solution, but also in the living cell. Real-time studies of fluorescent particles in open, laser-illuminated volume elements to unravel underlying inter- and intramolecular processes on time scales from nanoseconds to seconds, but also to uncover static and dynamic heterogeneities, i.e. differences in the molecular properties within ensembles of supposedly identical particles. Design of microfluidic systems for single particle manipulation.
Heinze KG, Koltermann A, and **Schwille P** (2000). Simultaneous Two-Photon Excitation of Distinct Labels For Dual-Color Fluorescence Cross-Correlation Analysis. *PNAS* **97**,10377-10382
Bacia K, Majoul IV, and **Schwille P** (2002). Probing the Endocytic Pathway in Live Cells Using Dual-Color Fluorescence Cross-Correlation Analysis. *Biophys. J.* **83**,1184-1193.

Date submitted: 23rd August 2002

Ine Segers-Nolten, (Ph.D. Student)

University of Twente, Applied Physics Dept.,
Biophysical Engineering Group,
PO Box 217,
7500 AE Enschede, the Netherlands.
Tel: +31 53 4893358   Fax: +31 4891105
g.m.j.segers-nolten@tn.utwente.nl
tnweb.tn.utwente.nl/bft/

Specialty Keywords: Single molecule fluorescence, Confocal, NER.

Scanning confocal fluorescence microscopy is used for a single molecule study of the Nucleotide Excision Repair process. NER-GFP fusion proteins are combined with fluorescently labeled DNA substrates to form complexes. Samples are prepared in agarose gel matrices, where uncomplexed DNA is rapidly diffusing and DNA-protein complexes are immobilized. Colocalization of GFP-label on the NER-protein with the DNA-label is an indication of complex formation. This method allows the study of protein-DNA binding under equilibrium conditions.
G.M.J. Segers-Nolten, C. Wyman, N. Wijgers, W. Vermeulen, A.T.M. Lenferink, J.H.J. Hoeijmakers, J. Greve, C. Otto, Scanning Confocal Fluorescence Microscopy for Single Molecule Analysis of Nucleotide Excision Repair Complexes, submitted to NAR, 2002.

Date submitted: 21st August 2002

Claus A. M. Seidel, Ph.D.

Heinrich-Heine- Universtaet Duesseldorf,
Institut fuer Physikalische Chemie,
Universitaetsstr. 1 Geb. 26 32 02
40225 Duesseldorf.
Tel: +49 211 81 14755 or 15284   Fax: +49 211 81 12803
Max-Planck-Institut fuer Biophysikalische Chemie, Abt. Fuer
Spektroskkopie und Photochemische Kinetik, Am Fassberg 11,
D 37077 Goettingen.
Tel: +49 551 201 1774 Fax: +49 551 2011501
cseidel@gwdg.de
www.mpibpc.gwdg.de/abteilungen/010/seidel/

Specialty Keywords: Single-molecule fluorescence spectroscopy, Multiparameter fluorescence detection (MFD).

It is my goal to obtain all information in a single-molecule experiment for applications in analytics and biophysics. Thus, as many fluorescence photons as possible must be detected, and a full set of fluorescence parameters must be registered by MFD: intensity, F, lifetime, *tau* and anisotropy, r, in several spectral windows together with its time-dependence [1].

[1] R. Kuehnemuth, C. A. M. Seidel; (2001) Principles of single molecule multiparameter fluorescence spectroscopy *Single Molecules* **2,** 251-254.

Date submitted: 26th August 2002

Paul R. Selvin, Ph.D.

University of Illinois at Urbana-Champaign,
Department of Physics, 363 LLP, MC704,
1110 W. Green Street; Urbana,
IL 61801-3080, USA.
Tel: (217) 244 3371   Fax: (217) 244 7187
selvin@uiuc.edu
www.physics.uiuc.edu/people/faculty/Selvin/

Specialty Keywords: Lanthanide luminescence, FRET, Single-molecule.

We develop and use fluorescence techniques with (sub)nanometer resolution, including new forms of FRET – e.g. single-pair FRET, FRET using luminescent lanthanide chelates. A major emphasis is developing new lanthanide chelates. Applications include measuring conformational changes in myosin and ion channels.

Cha, A., G. E. Snyder, P. R. Selvin, and F. Bezanilla. 1999. Atomic scale movement of the voltage sensing region in a potassium channel measured via spectroscopy. *Nature.* 402:809-813.

Selvin, P. R. 2002. Principles and Biophysical Applications of Luminescent Lanthanide Probes. *Annual Review of Biophysics and Biomolecular Structure.* 31:275-302.

Date submitted: 6th September 2002

Claudio H. Sibata, Ph.D.

Radiation Oncology, ECU School of Medicine,
600 Moye Blvd, Greenville,
Pitt County, 27858 4345,
USA.
Tel: (252) 816 2900   Fax: (252) 816 2812
sibatac@mail.ecu.edu
www.ecu.edu/radonc

Specialty Keywords: Cancer, Photodynamic therapy, Optical biopsy.

Research interests are optical biopsy, photodynamic therapy optimization for oncological patients, refine and improve therapy both by clinical modifications and via dosimetry enhancement.

Sibata CH, Colussi VC, Oleinick NL, Kinsella TK. Photodynamic Therapy in Oncology. Expert Opinion on Pharmacotherapy, 2001, 2:917-928.

Sibata CH, Colussi V, Oleinick NO, Kinsella TJ: Photodynamic Therapy: A New Concept in Medical Treatment. Braz J Med Biol Res, 2000, 33 :869-80

Date submitted: 22nd August 2002

Reiner Siebert, M.D.

Institute of Human Genetics, University Hospital Kiel,
Schwanenweg 24, Kiel,
Schleswig-Holstein, 24105,
Germany.
Tel: +49 431 597 1779   Fax: +49 431 597 1880
rsiebert@medgen.uni-kiel.de
www.uni-kiel.de/medgen

Specialty Keywords: Combined immunofluorescence and fluorescence in situ hybridization (FICTION).

With regard to fluorescence microscopy, the projects of our research group focus on the development of fluorescence in situ hybridization (FISH) assays for the detection of chromosomal abnormalities in tumors, as well as on the technical improvements of combined fluorescence immunophenotyping and interphase cytogenetics (FICTION technique). Current interests are the development of an automated platform for the detection of rare tumor cells and spot counting of multicolor hybridization signals.

J.I. Martin-Subero, I. Chudoba, L. Harder, S. Gesk, W. Grote, F.J. Novo, M.J. Calasanz, R. Siebert (2002). Multicolor-FICTION: Expanding the Possibilities of Combined Morphologic, Immunophenotypic, and Genetic Single Cell Analyses, *Am. J. Pathol.*, **161**, 413-420.

Date submitted: 10th September 2002

Aleksander Siemiarczuk, Ph.D.

Photon Technology International,
347 Consortium Court,
London, Ontario,
Canada N6E 2S8.
Tel: (519) 668 6920
asiemiarczuk@pti-can.com
www.pti-nj.com

Specialty Keywords: Time-resolved fluorescence, Lifetime distributions, TICT states.

Past and present research activities include co-discovery of Twisted Intramolecular Charge Transfer States (TICT), studies of intramolecular and solvation dynamics, long-range electron transfer in linked porphyrin-quinone derivatives, fluorescence in heterogeneous systems with lifetime distributions, development of a new methodology to study polydispersity of micelles using lifetime distributions, complexes with cyclodextrins, time-resolved fluorescence of proteins, photophysics of curcumin derivatives, development of time-resolved instrumentation.

A Time-Resolved and Steady-State Fluorescence Quenching Study on Naproxen and Its Cyclodextrin Complexes in Water, N. Sadlej-Sosnowska and A. Siemiarczuk (2001) Photochem. Photobiol. 138, 34-40.

Date submitted: 13th September 2002

Manoj K. Singh, Ph.D.

Department of Chemistry , University of Kansas,
1251 Wescoe Hall Drive, Lawrence,
KS, 66045, USA.
Tel: (785) 864 3679
mksingh@ku.edu & k_singh@vsnl.net
On leave from: Spectroscopy Division, B.A.R.C.
Mumbai-400085, India.

Specialty Keywords: Time-resolved spectroscopy, Single molecule fluorescence.

I have been mainly involved with the investigations on the photophysics, photochemistry and rotational dynamics of dye molecules in liquid phase using time-resolved fluorescence and transient absorption techniques. Recently, the focus of my research mostly involves the study of protein dynamics at the single molecule level. We are investigating the dynamics of calmodulin, a calcium signaling protein and its targets using single molecule time-resolved fluorescence and fluorescence polarization techniques.

1. M. K. Singh (2000) *Photochem. Photobiol.* **72**, 438-443.

Date submitted: 26th August 2002

Harald H. Sitte, M.D.

Molecular Neuropharmacology Group,
Institute of Pharmacology, University of Copenhagen,
Blegdamsvej 3, Copenhagen,
DK-2200 Copenhagen N, Denmark.
Tel: +45 3532 7549   Fax: +45 3532 7555
harald.sitte@univie.ac.at

Specialty Keywords: Fluorescence Microscopy, Fluorescence Resonance Energy Transfer, Membrane Proteins.

My research focuses on the understanding of the quaternary structure of membrane proteins, *i.e.* transport proteins like the serotonin or the GABA transporter. We use Fluorescence Resonance Energy Transfer Microscopy to learn more about their structural constraints and the impact, oligomerization may have on the function of these proteins.

Schmid, J. A., Scholze, P., Kudlacek, O., Freissmuth, M., Singer, E. A., and Sitte, H. H. (2001) Oligomerization of the Human Serotonin Transporter and of the Rat GABA Transporter 1 Visualized by Fluorescence Resonance Energy Transfer Microscopy in Living Cells *J.Biol.Chem.* **276,** 3805-3810.

Date submitted: 9th September 2002

Aleksandr V. Smirnov, Ph.D.

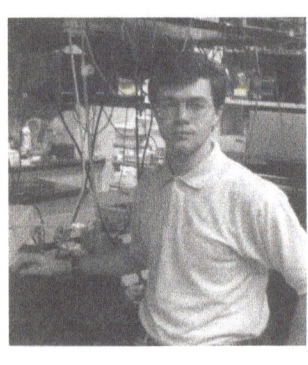

Laboratory of Biophysical Chemistry,
National Heart, Lung and Blood Inst., National Inst. of Health
NHLBI / NIH, Bldg.10, Rm. 5D14, 10 Center Drive,
MSC 1412, Bethesda, 20892, USA.
Tel: (301) 496 2558   Fax: (301) 480 6964
avs@helix.nih.gov
www.nhlbi.nih.gov/labs/biophysicalchem/index.htm

Specialty Keywords: Transient spectroscopy, Lasers, Biophysics.

My research interests focus on dynamical aspect of bimolecular function. The knowledge of structure and composition is essential but true understanding of mechanism involved is often impossible without direct observation of how it happens in real time. My methods of choice are femtosecond transient absorbance and laser induced fluorescence spectroscopy. This enables one to follow changes in state and environment of synthetic and natural optical probes, such as tryptophan. Also I develop stopped-flow techniques to study kinetics of biochemical reactions.

A. V. Smirnov *et. al.* (1997). Photophysics and Biological Applications of 7-Azaindole and its Analogs. *J. Phys. Chem. B*, **101**(15), 2758-2769.

Date submitted: 8th August 2002

Clint B. Smith, M.S.

U.S. Army Engineering Research and Development Center,
Fluorescence Remote Sensing Lab,
USAERDC-TEC, 7701 Telegraph Road, Alexandria,
Virginia, 22315, USA.
Tel: 703 428 8203   Fax: 703 428 8176
clint.b.smith@erdc.usace.army.mil
www.tec.army.mil

Specialty Keywords: Fluorescence Remote Sensing, Enzyme Substrates, Waterborne pathogens.

Research in the laboratory involves fluorescence remote sensing applications for waterborne pathogens. Novel fluorescent probes and enzyme substrates are utilized for the detection of pathogens existing in waterways using state-of-the-art fluorescent spectrometers. Applications are geared toward the imaging domain and will be developed after performing successful laboratory experiments binding molecular probes to specific targets.

Anderson, J.E., Webb, S.R., Fischer, R.L., Smith, C.B., Dennis, J.R., and Di Benedetto, J. (2002). *In Situ* Detection of the Pathogen Indicator *E.coli* Using Active Laser-Induced Fluorescence Imaging and Defined Substrate Conversion. *Journal of Fluorescence*. (12) 1 p. 51-55.

Date submitted: 22nd August 2002

Trevor A. Smith, Ph.D.

School of Chemistry,
University of Melbourne,
Victoria, 3010,
Australia.
Tel: +61 (3) 8344 6272   Fax: +61 (3) 9347 5180
trevoras@unimelb.edu.au
www.chemistry.unimelb.edu.au/research/groups/Photophysics/
photophysics

Specialty Keywords: Time-resolved fluorescence, Anisotropy, Microscopy.

Research Interests: Ultrafast laser spectroscopic techniques applied to photophysical processes in macromolecules such as polymers, photo-induced electron and energy transfer in supramolecules. Time resolved fluorescence microscopy techniques including multi-photon and confocal fluorescence microscopy. Time-resolved fluorescence anisotropy measurements, rheo-optical studies, time-resolved evanescent wave-induced fluorescence techniques.

T.A. Smith et al., (1998) "Time-Resolved Fluorescence Anisotropy Measurements of the Adsorption of Rhodamine-B Onto Colloidal Silica", *Coll. Polym.Sci.*, **276**, 1032-1037.
T.A. Smith, L.M. Bajada and D.E. Dunstan, (2002), "Fluorescence Polarization Measurements of the Local Viscosity of Hydroxypropyl Guar in Solution", *Macromolecules*, **35**, 2736-2742.

Date submitted: 1st May 2002

Peter, T. C. So, Ph.D.

Department of Mechanical Engineering,
Department of Biological Engineering,
Massachusetts Institute of Technology,
3-461, 77 Mass Ave,
Cambridge, MA, 02139,
USA.
Tel: (617) 253 6552   Fax: (617) 258 9346
ptso@mit.edu

Specialty Keywords: Multi-photon microscopy, Time-resolved spectroscopy, Correlation spectroscopy.

My research focuses on the development of instrumentation for biomedical studies. Recent projects in my laboratory include video rate two-photon microscopy, fluorescence correlation spectroscopy, 3-D image cytometery. These instruments are applied in studies such as: protein dynamics, cellular mechanotransduction, tissue carcinogenesis, and non-invasive optical biopsy.

So et al., "Two-Photon Excitation Microscopy", Annu. Rev. Biomed. Eng., 2, 399-429 (2001).
Huang et al., "Three-Dimensional Cellular Deformation Analysis with a Two-Photon Magnetic Manipulator Workstation," Biophys. J., 82, 2211-2223 (2002).

Date submitted: 27th August 2002

Steven A. Soper, Ph.D.

Department of Chemistry, Louisiana State University,
232 Choppin Hall, Baton Rouge, LA,
70803-1804,
USA.
Tel: (225) 578 1527   Fax: (225) 578 3458
chsope@lsu.edu
www.cmm.lsu.edu

Specialty Keywords: Near-IR Fluorescence, Time-Resolved Fluorescence, Single Molecule Detection.

Ultrasensitive time-resolved fluorescence spectroscopy; dye photophysics and photochemistry; bioanalytical and environmental applications of near-infrared fluorescence; capillary zone and gel electrophoresis using fluorescence detection; development of novel laser-based DNA analysis schemes; bioanalytical applications of laser-induced fluorescence detection; development of microfabricated biochemical analysis systems; single molecule detection using near-IR fluorescence detection.

E. Waddell, Y. Wang, W. Stryjewski, S. McWhorter, A. Henry, D. Evans, R. L. McCarley and **S. A. Soper**, *Anal. Chem.* 72 (2000) 5907.
S. Lassiter and **S.A. Soper,** *Electrophoresis* 23 (2002) 1480.

Date submitted: 13th September 2002 **Ian Soutar, Ph.D.**

ChemistryDepartment,
University of Sheffield,
Brook Hill, Sheffield,
S3 7HF, UK.
Tel: +44 (0)114 222 9561   Fax: +44 (0)114 273 8673
i.soutar@sheffield.ac.uk

Specialty Keywords: Anisotropy, Energy Harvesting, Polymers.

Research Interests: Studies of polymer behavior both in solution and the solid state using time-resolved emission anisotropy. Water-soluble polymers. Smart systems. Polymers for energy harvesting and solar energy conversion.

D. Allsop, L. Swanson, I. Soutar et al. (2001) "Fluorescence Anisotropy: A Method for Early Detection of Alzheimer β-Peptide Aggregation" *Biochem. Biophys. Res. Comm.,* **285**, 58-63.
C.K. Chee, I. Soutar et al. (2001) "Time-resolved Fluorescence Studies of the Interactions Between the Thermoresponsive host, PNIPAM, and Pyrene" *Polymer,* **42**, 1067-1071.

Date submitted: 23 August 2002 **Elias Stathatos, Ph.D.**

Engineering Science Dept,
University of Patras,
26500 Patras,
Greece.
Tel: 30 610997587   Fax: 30 610997803
stathatos@des.upatras.gr

Specialty Keywords: Photophysics, Sol-gel, Solid-state electrolytes.

Research interests include steady-state and time-resolved fluorescence characterization of nanocomposite thin films and transparent solid matrices. Applications involve dye-sensitized photoelectrochemical cells, photocatalytic metal oxide surfaces, lasing in nanocomposite and organic materials and electroluminescence of ligand lanthanide complexes.

E.Stathatos, P.Lianos, Ch.Krontiras (2001) *J.Phys.Chem. B. 105, 3486-3492.*
E.Stathatos, P.Lianos, U.Lavrencic Stangar and B. Orel. (2001) *Chem. Phys. Letters* 345, 381-385.

Date submitted: 2nd August 2002

Lorenzo Stella, Ph.D.

Department of Chemical Sciences and Technologies,
University of Roma Tor Vergata,
Via della ricerca scientifica, 00133, Roma,
Italy.
Tel: +39 06 7259 4463   Fax: +39 06 7259 4328
stella@stc.uniroma2.it

Specialty Keywords: Peptide, protein structure and dynamics,
Peptide-membrane interactions.

My main research focus is the application of fluorescence spectroscopy in the study of protein and peptide structure and dynamics. Some current research projects: mechanism of action of antimicrobial peptides and their interaction with membranes; design and characterization of peptide-based molecular devices utilizing photophysical processes for memories, switches and energy conversion; Comparisons between time-resolved fluorescence spectroscopy and computer simulations.

L. Stella et al. (2002) Structural features of model glycopeptides in solution and in membrane phase. A spectroscopic and molecular mechanics investigation. *Biopolymers* **64**, 44-56.

L. Stella (2001) Comparisons between time-resolved fluorescence experiments and computer simulations. In "Spectroscopic techniques in biophysics", IOS Press (Amsterdam), pp. 89-103.

Date submitted: 11th September 2002

Daniel W. Stockholm, Ph.D.

Laboratoire d'imagerie, Genethon,
1 bis rue de l'Internationale, Evry,
91000,
France.
Tel: 01 60 77 86 98
stockho@genethon.fr

Specialty Keywords: Confocal microscopy, Muscle
visualization, Real-time PCR.

We are part of a research center focussed on gene therapy and run a core service for imaging with 2 confocal microscopes including a multiphon. We use FRET for the study of calpain function and are developing techniques for the intra vital imaging. We also acquired some expertise in real-time PCR and use it extensively for gene expression studies and viral titration.

Stockholm D, et al. , *Am J Physiol Cell Physiol,* 2001, Jun;280(6):C1561-9.

Feasson L, Stockholm D, et al. *J Physiol.* 2002 Aug 15;543(Pt 1):297-306.

Date submitted: 22nd August 2002

John C. Sutherland, Ph.D.

Biology Department,
Brookhaven National Laboratory,
Upton, NY, 11973,
USA.
Tel: 631 344 3279
jcs@bnl.gov
bnlstb.bio.bnl.gov/biodocs/structure/J_Sutherland.htmlx

Specialty Keywords: Time-resolved fluorescence and circular dichroism using synchrotron radiation. DNA damage quantitation by gel electrophoresis and single molecule sizing.

Pioneered the use of synchrotron radiation for the measurement of circular dichroism and time-resolved fluorescence spectroscopy in the ultraviolet/visible spectral regions. Invented the Fluorescence Omnylizer, a single-photon counting detector that records the time-delay, wavelength and polarization of each detected photon. First to use CCD camera to record image of DNA fluorescence in electrophoretic gels. Uses gel fluorescence or single-molecule laser fluorescence sizing to quantify DNA damage by average length analysis.

Date submitted: 13th September 2002

Linda Swanson, Ph.D.

ChemistryDepartment,
University of Sheffield,
Brook Hill, Sheffield,
S3 7HF, UK.
Tel: +44 (0)114 222 9564 Fax: +44 (0)114 273 8673
l.swanson@sheffield.ac.uk

Specialty Keywords: Anisotropy, Smart Polymers, Polymer Dynamics.

Research Interests: Anisotropy studies of the conformational behavior of smart polymers. Polymer dynamics. Polymer interactions (synthetic and biomacromolecules). Polymer relaxation behavior in the solid state. Novel polymeric materials for enhanced solar energy conversion.

L. Swanson, et al., (2001). "Manipulating the thermoresponsive behaviorof PNIPAM "*Macromolecules* **34**, 544-754 .
N. J. Flint, S. Gardebrecht and L. Swanson, (1998). "Luminescence investigations of smart microgel systems", *J. Fluorescence,* **8**, 343-353.

Date submitted: 10th September 2002

Kerry M. Swift, M.S.

Abbott Laboratories,
Global Pharmaceutical Research and Development,
Department of Structural Biology R46Y / AP9LL,
Abbott Park, IL 60064-6114, USA.
Tel: 847 937 7289   Fax: 847 935 0143
Kerry.Swift@abbott.com

Specialty Keywords: Drug discovery, Binding, Fluorescence lifetimes, FCS, HTS.

My research within the optical spectroscopy group here at Abbott in the last 10 years has been toward characterizing or improving fluorescent probe-based assays for testing of drug-like compounds. Furthermore, I sometimes use the intrinsic fluorescence of proteins to study their structure or binding. I am also developing the use of Raman microscopy on protein crystals.

Sergey Y. Tetin, Kerry M. Swift and Edmund D. Matayoshi (2002). Measuring antibody affinity and performing immunoassay at the single molecule level *Analytical Biochemistry* **307**(1) 84-91.
A. M. Petros, A. Medek, D. G. Nettesheim, D. H. Kim, H. S. Yoon, K. Swift, E. D. Matayoshi, T. Oltersdorf and S. W. Fesik (2001). Solution structure of the anti-apoptotic protein bcl-2 *Proc. Natl. Acad. of Sci. USA* **98**(6) 3012-3017.

Date submitted: 13th September 2002

Henryk Szmacinski, Ph.D.

Microcosm, Inc.,
9140 Guilford Road, Suite O,
Columbia, MD21046,
USA.
Tel: (301)725 2775   Fax: (301)725 2941
henryks@microcosm.com

Specialty Keywords: Spectroscopy, Fluorescence Probes, Optical Sensing.

My research interests include UV/VIS spectroscopy, optical sensors and biosensors, frequency-domain time resolved spectroscopy, and multi-photon microscopy. This involves of application of fluorescence lifetime to chemical sensing and imaging, immunoassays, DNA hybridization and cellular studies. Current interest is in development of disposable sensor arrays for biotechnology and clinical chemistry and exploring enhanced fluorescence using metallic nano-structures.
Measurement of Intensity of Long Lifetime Luminophres in the Presence of Background Signals Using Phase-Modulation Fluorometry. H. Szmacinski and J.R. Lakowicz, Appl. Specrosc. 53:1490-1495, 1999.

Date submitted: 3rd September 2002

Patrice Talaga, Ph.D.

Chemical Research, UCB S.A.,
Chemin du Foriest, Braine l'Alleud,
1420,
Belgium.
Tel: 32 2 386 2727   Fax: 32 2 386 2704
patrice.talaga@ucb-group.com

Specialty Keywords: Drug discovery, Medicinal chemistry, Alzheimer's Disease.

Current Interests: External Chemical Research. Management of Academic & CRO collaborations in Chemistry, Combichem & custom synthesis. Research interest in CNS (Alzheimer's Disease, Parkinson's Disease, and Epilepsy...) and Immuno-Allergy (Rhinitis, Asthma...) areas. Particular interest in Amyloid aggregation related research.

β-Amyloid Aggregation Inhibitors for the Treatment of Alzheimer's Disease: Dream or Reality? P. Talaga. Mini Reviews in Med. Chem. 2001, 1, 175-186.

First Dual NK1 Antagonists-Serotonin Reuptake Inhibitors: Synthesis and SAR of a New Class of Potential Antidepressants. T. Ryckmans et al. Bioorg. Med. Chem. Letters 2002, 12, 261-264..

Date submitted: 22nd July 2002

Fumio Tanaka, Ph.D.

Mie Prefectural College of Nursing,
Yumegaoka 1-1-1,
Tsu 514-0116,
Japan.
Tel / Fax: +81 59 233 5640
fumio.tanaka@mcn.ac.jp
www.nurse.mcn.ac.jp/

Specialty Keywords: Fluorescence, Flavin, Theory of Anisotropy.

Current Research Interests: I am working mostly on the time-resolved fluorescence of tryptophan and flavins in proteins in sub-picosecond region. I was much inspired on theory of anisotropy by knowing Weber's the Additivity Law of Polarization. I still have interested in developing the theory of fluorescence anisotropy.

N. Mataga et al.(2000). Dynamics and mechanisms of ultrafast fluorescence quenching reactions of flavin chromophores in protein nanospace: *J. Phy. Chem. B*, **104**, 10667-10677.

N. Tamai et al.(2002). Solvation dynamics of the excited 1,2-(p-cyano-p'-methoxydiphenyl)-ethyne: *J. Phys. Chem. A*, **106**, 2164-2172.

Date submitted: 26th July 2002

Hans J. Tanke, Ph.D.

Department Molecular Cell Biology,
Leiden University Medical Center,
Wassenaarseweg 72, 2333 AL Leiden,
The Netherlands.
Tel: +31 71 5276196   Fax: +31 71 5276180
H.J.Tanke@lumc.nl

Specialty Keywords: Fluorescence technology, Molecular analysis, Microscopy.

The study of the molecular composition of cells and chromosomes, using fluorescence labeling technology (FISH, immunocytochemistry, GFP) and (automated) digital microscopy, in order to unravel the molecular mechanisms that determine normal and abnormal cell function. The use of this information and methodology to develop improved diagnostic methods to be applied in the field of genetics, haematology and oncology.

Rijke F.v.d. et al. Up-converting phoshor reporters for nucleic acid microarrays. Nature Biotechnology 19:273-276, 2001. Ref. 2: Snaar SP et al. Mutational analysis of fibrillarin and its mobility in living cells. J. Cell Biol. 151:653-662, 2000.

Date submitted: 13th September 2002

Olga Tchaikovskaya, Ph.D.

Department of Photonic Complex Molecules,
Siberian Physical Technical Institute,
1, Novo-Sobornaia sq., Tomsk, 634055,
Russia.
Tel: +7 3822 533426   Fax: +7 3822 53 3034
tchon@phys.tsu.ru

Specialty Keywords: Photophysics, Photochemistry.

The photophysical and photochemical properties of phenols in aqueous solution with and without irradiation were investigated[1]. Also the phenols photolysis in water with humic acid, a comparative analysis of the efficiency of photochemical and microbiological phenol destruction were studied[2].

O.N.Tchaikovskaia, I.V.Sokolova, V.A.Svetlichnyi, et al. Journal of Fluorescence, Vol. 10, No. 4, 2000, P. 403-408.

O.N.Tchaikovskaia, I.Sokolova, L.Kondratieva, et al. Inter. J. of Photoenergy, 2001, Vol.3, No.4, P.177-180.

Date submitted: 24 July 2002

Richard B. Thompson, Ph.D.

Department of Biochemsitry and Molecular Biology,
University of Maryland School of Medicine,
108 N. Greene Street, Baltimore,
Maryland 21201, USA.
Tel: (410) 706 7142   Fax: (410) 706 7122
rthompso@umaryland.edu

Specialty Keywords: Biosensors, Fiber optic sensors, Metal ions
"Our work has focused on fluorescence-based biosensors and fiber optic biosensors. Our metal ion biosensors employ carbonic anhydrase II variants as recognition molecules, which transduce the concentrations of metal ions such as Cu(II), Zn(II), and others as changes in fluorescence lifetime, polarization, or intensity ratio. Carbonic anhydrase gives the sensor unmatched sensitivity (to picomolar and below) and selectivity (demonstrated in sea water and cerebrospinal fluid), which both can be modulated by subtle changes in the protein structure. Use of optical fiber permits remote, continuous monitoring in situ.

C. A. Fierke and R. B. Thompson, "Fluorescence-based biosensing of zinc using carbonic anhydrase," BioMetals 14 (3-4) 205-222 (2001).

R. B. Thompson, et al., "Fluorescent zinc indicators for neurobiology," Journal of Neuroscience Methods 118, 63-75 (2002).

Date submitted: 6th May 2002

Leann Tilley, Ph.D.

Department of Biochemistry, La Trobe University,
Plenty Rd,
Bundoora, Melbourne 3086,
Australia.
Tel: 61 3 94791375
L.Tilley@latrobe.edu.au
www.latrobe.edu.au/biochemistry/

Specialty Keywords: FRAP, Far red fluorophore, GFP.

Use of fluorescence recovery after photobleaching protocols and GFP transfection to study the molecular dynamics of proteins in uninfected and malaria parasite-infected erythrocytes. Synthesis and characterisation of novel far red-absorbing chromophores.

Klonis, N., Rug, M., Wickham, M., Harper, I., Cowman, A. and Tilley, L. (2002) Fluorescence photobleaching analysis for the study of cellular dynamics. *European Journal of Biophysics* (review), 2002, 31, 36-51.

Klonis, N., Wang, H., Quazi, N.H., Casey, J.L., Neumann, G., Deady, L.W. and Tilley, L. (2001) Characterisation of a Series of Far Red Absorbing Perylene Diones: A New Class of Fluorescent Probes for Biological Applications. *Journal of Fluorescence* 11, 1-11.

Date submitted: 13th September 2002

Ferenc G. Tölgyesi, Ph.D.

Dept. of Biophysics and Radiation Biology,
Semmelweis University,
Puskin u. 9, Budapest,
H-1088, Hungary.
Tel: (36 1) 266 2755 / 4033   Fax: (36 1) 266 6656
tolgyesi@puskin.sote.hu
www.biofiz.sote.hu

Specialty Keywords: Tryptophan phosphorescence.

Research interests: protein structure and dynamics, their relation to function; small heat shock proteins; protein aggregation; effect of high pressure on proteins; luminescence spectroscopy, tryptophan phosphorescence, absorption spectroscopy.

Tölgyesi, F., Ullrich, B., Fidy J (1999) Tryptophan phosphorescence signals characteristic changes in protein dynamics at physiological temperatures *Biochim. Biophys. Acta*, **1435**, 1-6.
Ullrich B., Laberge M., Tölgyesi F., Szeltner Z., Polgár L., Fidy J. (2000) Trp 42 rotamers report reduced flexibility when the inhibitor acetyl pepstatin is bound to HIV -1 protease *Protein Science.* **9**, 1-14.

Date submitted: 13th September 2002

John M. Torkelson, Ph.D.

Dept. of Chemical Engineering,
Dept of Materials Science & Engineering,
Northwestern University, 2145 Sheridan Road,
Evanston, IL 60208-3120, USA.
Tel: 847 491 7449
j-torkelson@northwestern.edu

Specialty Keywords: Polymers, Thin Films, Dynamics.

Luminescence studies of polymers related to reactive processing, gelation, relaxation processes, diffusion and diffusion-controlled reactions, and nanoscale confinement.

S.D. Kim and J.M. Torkelson (2002). Nanoscale confinement and temperature effects on associative polymers in thin films: Fluorescence study of a telechelic, pyrene-labeled polydimethylsiloxane. *Macromolecules* **35**, 5943-5952.

C.J. Ellison, S.D. Kim, D.B. Hall and J.M. Torkelson (2002). Confinement and processing effects on glass transition and physical aging in ultrathin polymer films: Novel fluorescence measurements *Eur. Phys. J. E* **8**, 155-156.

Date submitted: 19th September 2002

Eric Trinquet, M.Sc.

HTRF Research, Cis Bio International,
B.P. 84175,
30204 Bagnols sur Ceze Cedex,
France.
Tel: 33 (0)4 66 79 67 69 Fax: 33 (0)4 66 79 19 20
etrinquet@cisbiointernational.fr
www.htrf-assays.com

Specialty Keywords: Rare Earth Cryptates, FRET, Biomolecular Interactions.

Fields of interest: Research on Fluorescence based techniques for probe molecular interactions probing. Research on new methods based on the use of FRET combined with long lived fluorophores. Applications in High Troughtput Screening, Cellular Biology and Molecular Biology.

H. Bazin, E. Trinquet, G. Mathis (2002). Time Resolved Amplification of Cryptate Emission: a Versatile Technology to trace Biomolecular Interactions. Review in Molecular Biotechnology, **82**,233-250.

E. Trinquet, F. Maurin, M. Préaudat, G. Mathis (2001) Allophycocyanin 1 as a near infrared fluorescent tracer: isolation, characterization, chemical modification and use in homogeneous fluorescence resonance energy transfer system. Analytical Biochemistry, **296**, 232-244.

Date submitted: 28th August 2002

Acuña A. Ulises, Ph.D.

Department of Biophysics,
Instituto de Química-Física "Rocasolano", C.S.I.C,
119 Serrano, 28006-Madrid,
Spain.
Tel: +34 91 561 9400 Fax: +34 91 564 2431
roculises@iqfr.csic.es

Specialty Keywords: Polarised luminescence spectroscopy, Fluorescent bioprobes, Membrane structure and dynamics.

Research fields: Probing protein and lipid membrane dynamics with time-resolved fluorescence, phosphorescence and T-T dichroism. Synthesis of new fluorescent labels and probes. Theory of rotational depolarisation of luminescence. Fundamental photochemistry: triplet-triplet energy transfer end excited-state proton transfer. The history of solution fluorescence.

M.L.Ferrer, R.Duchowicz, B.Carrasco, J.Garcia de la Torre and A.U. Acuña (2001). The conformation of serum albumin in solution *Biophysical J.* **80**, 2422-2430.

E. Quesada, A.U. Acuña and F. Amat-Guerri (2001) New transmembrane polyene bolaamphiphiles as fluorescent probes in lipid bilayers *Angew, Chem. Int. Ed.* **40**, 2095-2097.

Vaganova, E.
Valenta, J.

Date submitted: 29th August 2002

Evgenia Vaganova, Ph.D.

The Institute of Chemistry,
The Hebrew University of Jerusalem,
Givat Ram, Jerusalem,
94904, Israel.
Tel: 972 2 658 4199
gv@cc.huji.ac.il

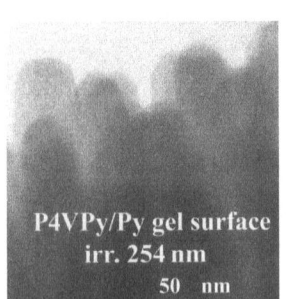

Specialty Keywords: Photochemistry, Gel, Pyridine.

The mixture of Poly(4-vinyl pyridine)/pyridine is a novel photosensitive matter [1]. Depending on the irradiation wavelength different gel's structures are formed. Different emitting centers are in correlation with photoinduced structures. The photosensitivity of the composition is based on the formation of the two-molecular structure resulted from the interaction of self-protonated polymeric pyridinium ion with free pyridine. Open form photoproduct (irradiation at 250 nm), proton shuttle (irradiation at 380 nm) [2] are responsible for the different photoinduced gel's formation.

E. Vaganova, G. Meshulam, et. al (2000) *J.of Fluorescence* **10**, 81--89.
E. Vaganova, V. Hodorokovsky, L.Filatov, and S. Yitzchaik (2000) *Adv. Materials* **12**, 1679--1671.

Date submitted: 30th August 2002

Jan Valenta, Ph.D.

Department of Chemical Physics & Optics,
Faculty of Mathematics & Physics Charles University,
Ke Karlovu 3,
CZ-121 16 Prague 2, Czech Republic.
Tel: +420 2 2191 1272 Fax: +420 2 2191 1249
jan.valenta@mff.cuni.cz
kchf-45.karlov.mff.cuni.cz/~JanValenta

Specialty Keywords: Single-nanocrystals, Luminescence.

Optical spectroscopy of individual low-dimensional semiconductor structures (nanocrystals - quantum dots) and biological complexes.
Linear and non-linear optical properties of semiconductors and insulators (pump-and-probe techniques, four-wave-mixing, transient and persistent spectral hole-burning and hole-filling).

J. Valenta, R. Juhasz, and J. Linnros: Optical spectroscopy of single silicon quantum dots (2002) *Appl. Phys. Lett.* **80** (6), 1070-1072.

J. Valenta, J. Dian, J. Hála, P. Gilliot, and R. Lévy: Persistent spectral hole-burning and hole-filling in CuBr semiconductor nanocrystals (1999) *J. Chem. Phys.* **111**, 9398-9405.

Date submitted: 2nd May 2002

Bernard Valeur, Ph.D.

Conservatoire National des Arts et Métiers,
292, rue Saint-Martin,
75141 Paris Cedex 03,
France.
Tel: +33 (0)1 40 27 26 22   Fax: +33 (0)1 40 27 23 62
valeur@cnam.fr
Specialty Keywords: Fluorescent molecular sensors. Excitation
energy transfer Multichromophoric systems.

Current interests: Design of fluorescent molecular sensors for ion recognition (e.g. calixarene-based fluorescent sensors for the detection of alkali, alkaline-earth and heavy metal ions). Multichromophoric systems (e.g. antenna effect and energy hopping in multichromophoric cyclodextrins; multichromophoric calixarenes for ion detection; excitation energy transfer in porphyrin assemblies). Investigation of surfaces by fluorescence spectroscopy (e.g. characterization of the distribution of hydroxyl groups on alumina surfaces via excimer formation of grafted pyrene probes).
B. Valeur (2002). Molecular Fluorescence. Principles and Applications. Wiley-VCH, Weinheim.
B. Valeur and I. Leray (2000). Design principles of fluorescent molecular sensors for cation recognition, *Coord. Chem. Rev.* **205**, 3-40.

Date submitted: 3rd September 2002

German-Dutch Wind Tunnels

Reinier K. van der Draai,

Data and Controls, German-Dutch Wind Tunnels (DNW),
Anthony Fokkerweg 2,
Amsterdam, 1059 CM,
The Netherlands.
Tel: +31 020 5113382   Fax: +31 020 5113131
draai@dnw.aero
www.dnw.aero

Specialty Keywords: PSP, Decay, Intensity.

A single luminophor Rhutinium paint was developed to be able to measure the pressure at the surface of an aircraft model. The methods used for this technique were decay and intensity.
The decay method still needs improvement. The intensity method is presently installed in the windtunnel.

Date submitted: 28th August 2002

Kelly A. Van Houten, Ph.D.

Sensors for Medicine and Science Inc,
12321 Middlebrook Road, Ste. 210,
Germantown, MD 20874,
USA.
Tel: (301) 515 7260   Fax: (301) 515 0988
kvanh@s4ms.com
www.s4ms.com

Specialty Keywords: Glucose sensor, Dual-emitter, Oxygen sensor.

Currently, I am working on a fluorescence-based sensor for continuous in-vivo glucose monitoring. My interests involve the design of novel dual-emitting metal complexes as sensors and probes.

Van Houten, K.A.; Pilato, R.S. (1999) in K.S. Schanze; V. Ramamurthy (Eds.) *Molecular and Supramolecular Photochemistry*: *Multimetallic and Macromolecular Inorganic Chemistry*, Marcel Dekker Inc., New York, pp. 185-214

Van Houten, K.A.; Heath, D.C.; Pilato, R.S. (1998) *J. Am. Chem. Soc.* **120**, 12359.

Date submitted: 29th August 2002

Martin J. vandeVen, Ph.D.

Department MBW, Biomedical Research Institute (BIOMED) /
Institute of Materials Research (IMO),
Limburg University Center (LUC), Bldg D / Trans National,
University Limburg (tUL), University Campus,
Diepenbeek, B-3590, Belgium.
Tel: 0032 (0)11 268558 / 8816   Fax: 0032 (0)11 268599 / 8899
martin.vandeven@luc.ac.be
www.luc.ac.be/engels/ & www.tul.edu

Specialty Keywords: Spectrofluorimetry, Microscopy, Image analysis.

Collaborative research centers on (1) fluorescence imaging microscopy of cellular interactions in autoimmune diseases Multiple Sclerosis (MS) and Rheumatoid Arthritis (RA) (2) polymer fluorescence characterization for biosensors (3) Chlorophyll and GFP fluorescence imaging related to leaf and fruit physiology (4) application of neural networks in image analysis (5) development of laser-based time- and frequency domain fluorescence methodologies at the LUC Biomed fluorescence Center.

Using fluorescence images in classification of apples. Codrea, M.; Tyystjärvi, E.; vandeVen, M.; Valcke, R. and Nevalainen, O.; IASTED-VIIP Benalmadena, Malaga, Spain Sept. 9-12 2002.

Date submitted: 5ᵗʰ May 2002

Wilfried G.J.H.M. Van Sark, Ph.D.

Molecular Biophysics, Debye Institute, Utrecht University,
P.O. Box 80000, Utrecht,
NL-3508 TA,
The Netherlands.
Tel: +31 30 253 2825   Fax: +31 30 253 2706
vansark@phys.uu.nl
www.phys.uu.nl/~wwwmbf

Specialty Keywords: Single molecule imaging / spectroscopy, Quantum dots.

Initiate, coordinate and perform scientific research in the field of fast fluorescent imaging and spectroscopy of (single) semiconductor quantum dots, colloidal systems, and organic molecules. This includes bioconjugation and application in (artificial) membrane systems.

W. G. J. H. M. Van Sark, et al. (2002). Time-Resolved Fluorescence Spectroscopy Study on the PhotoPhysical Behaviour of Quantum Dots *J. Fluoresc.* **12,** 69-76.

W. G. J. H. M. Van Sark, et al. (2001). Photo-oxidation and Photobleaching of Single CdSe/ZnS Quantum Dots probed by Room-Temperature Time-Resolved Spectroscopy *J. Phys. Chem. B* **105,** 8281-8284.

Date submitted: 21ˢᵗ August 2002

David Vaudry, Ph.D.

European Institute for Peptide Research (IFRMP 23),
Laboratory of Cellular and Molecular Neuroendocrinology,
INSERM U413, UA CNRS,
University of Rouen76821 Mont-Saint-Aignan Cedex, France.
Tel: (33) 235 14 6760   Fax: (33) 235 14 6946
David.vaudry@univ-rouen.fr
www.univ-rouen.fr/inserm-u413/microscopie.htm

Specialty Keywords: Confocal microscopy, Microarray, Q-RT-PCR & Microplate reader.

We are studying the molecular and cellular mechanisms involved in the neurotrophic and antiapoptotic activities of the neuropeptide PACAP. The genes regulations and functions are investigated using the microarray, Q-RT-PCR or siRNA techniques. The protein levels and activities are measured by calcium imaging, western blotting or enzyme kinetics.

D. Vaudry et al. (2002) Pituitary adenylate cyclase-activating polypeptide protects rat cerebellar granule neurons against ethanol-induced apoptotic cell death. *Proc. Natl. Acad. Sci. USA* **99**: 6398-6403.

D. Vaudry et al. (in press) Analysis of the PC12 cell transcriptome after differentiation with pituitary adenylate cyclase-activating polypeptide (PACAP) *J. Neurochem.*

Date submitted: 9th September 2002

Jose Luis Vazquez-Ibar, Ph.D.

Howard Hughes Medical Inst. Dept. of Physiology and
Microbiology & Molecular Genetics, Molecular Biology Inst.,
University of California Los Angeles, 5748 MRL,
675 Charles E. Young Drive South,
Los Angeles CA 90095-1662.
Tel: (310) 206 5055
vazquez@hhmi.ucla.edu

Specialty Keywords: FRET, Lanthanide luminescence, Protein engineering.

Research focused on: study of structure and dynamics of membrane proteins by combining protein engineering and fluorescence techniques. In particular, we have developed a new approach to perform FRET measurements in an integral membrane protein using the luminescence of a lanthanide atom as energy donor. We created a terbium binding site in the middle cytoplasmic loop of lactose permease of E. coli (LacY) by engineering a Ca^{2+} coordinating sequence (EF-hand motif) with altered specificity for terbium.

Vazquez-Ibar JL., Weinglass, A.B. & Kaback, H. R. (2002) *Proc Natl Acad Sci USA* **99**, 3487-3492.

Date submitted: 23rd August 2002

Nikolai L. Vekshin, Ph.D.

Institute of Cell Biophysics,
Institutskaya street-3, Pushchino,
Moscow region, 142290,
Russia.
Tel: 923 74 67 2 92
nvekshin@rambler.ru
photonics.narod.ru

Specialty Keywords: Photonics, Biophysics, Spectroscopy.

Nikolai Vekshin has 5 books, 2 patents (multipass cuvettes for fluorescence) and many papers in international journals. His scientific interest is photophysics and spectroscopy of biopolymers. He uses: steady-state, synchroneous, polarization and time-resolved fluorescence methods, phosphorescence, IR spectroscopy, luminescence microscopy, etc. He developed a number of new methodical approaches for high-sensitive detection and investigations of proteins, nucleic acids and membranes. The main part of his job was concerned with fluorescence energy transfer. His work was supported by RFFI, NWO, NATO, FEBS, NSF, CRDF, and so on.
Vekshin N.L. Energy Transfer in Macromolecules. Bellingham, SPIE, 1997.
Vekshin N.L Photonics of Biopolymers. Springer, 2002.

Date submitted: 4th July 2002

Rance A. Velapoldi, Ph.D.

Nygaardskogen 28,
N-3408 Tranby,
Norway,

Tel: 047 32 853445
velapoldi@netcom.no

Specialty Keywords: Fluorescence standards, Corrected spectra, Quantum yields.

In late 60's and 70's, performed extensive research on organic species in solution and inorganic ion-doped glasses for use as macro- and micro-luminescence standards in addition to some analytical applications of fluorescence at the National Bureau of Standards, Washington, DC. (now NIST). Retired from NIST in 1999 but continuing research on standards and luminescence at the Pharmacy Institute, University of Oslo, Blindern, Norway.

R.A. Velapoldi and K.D. Mielenz, NBS Special Publication 260-64, US Department of Commerce, Washington, DC. 1980.

R.A. Velapoldi and M.S. Epstein, Luminescence Standards for Macro- and Microspectrofluorimetry; in "Luminescence Applications" M.C. Goldberg, Ed. ACS Symposium Series, 383, pp 98-126 (1989), Washington, DC.

Date submitted: 29th August 2002

Nel H. Velthorst, Ph.D.

Analytical Chemistry & Applied Spectroscopy, Laser Centre,
Vrije Universiteit Amsterdam,
de Boelelaan 1083, Amsterdam,
1081 HV, The Netherlands.
Tel: +31 20 4447541   Fax: +31 20 4447543
velthrst@chem.vu.nl
www.chem.vu.nl/acas/

Specialty Keywords: Laser induced and high-resolution molecular fluorescence.

The research has been directed on the potential of laser-induced fluorescence detection coupled to CE and LC and on the development and application of Shpol'skii Spectroscopy and Fluorescence Line Narrowing Spectroscopy for identification in analytical and environmental analysis, in particular applied on polycyclic aromatic hydrocarbons and their metabolites.

O.F.A. Larsen, I.S. Kozin, A.M. Rijs, G.J. Stroomberg, J.A. de Knecht, N.H. Velthorst and C. Gooijer: Direct identification of pyrene metabolites in organs of the isopod *Porcellio scaber* by Fluorescence Line Narrowing Spectroscopy. Anal. Chem. 70, 1182-1185 (1998)

Date submitted: 30th August 2002

Mariano Venanzi, Ph.D.

Department of Chemical Sciences and Technologies,
University of Roma Tor Vergata,
Via della ricerca scientifica, 00133, Roma,
Italy.
Tel: +39 06 7259 4468   Fax: +39 06 7259 4328
venanzi@uniroma2.it

Specialty Keywords: Biospectroscopy, Energy / electron transfer, Peptide structure.

My research actvity focusses on the application of fluorescence spectroscopy and other photophysical techniques in the study of energy/electron flow in peptides and molecules of biological interest. Current research projects: structure of peptide foldamers; design and characterization of peptide-based molecular devices for memories, switches and energy conversion; energy/electron transfer in porphyrin dimers and dendrimers; photocatalysis in micelles and organized environments.

(2002) Structural features and conformational equilibria of 3_{10}-helical peptides in solution by spectroscopic and molecular mechanics studies *Biopolymers(Biospectroscopy)* **67**, 247-250.

(2002) Effects of helical distortions on the optical properties of amide NH infrared absorption in short peptides in solution. *J. Phys. Chem. B* **106**, 5733-5738.

Date submitted: 29th July 2002

Jo Vercammen, Ph.D.

Biochemistry, K.U.Leuven,
Celestijnenlaan,
Heverlee, 3001,
Belgium.
Tel: 00 32 16 327132   Fax: 00 32 16 327982
Jo.Vercammen@fys.kuleuven.ac.be

Specialty Keywords: HIV-1 integrase, Fluorescence Correlation Spectroscopy, Fluorescence Fluctuation Analysis.

The Laboratory of Biomolecular Dynamics is equipped with an FCS instrument and within this project this technique will be developed for the study of the enzyme integrase. HIV-1 integrase is a lentiviral protein and is regarded as one of the potential candidates for developing antiviral drugs, next to reverse transcriptase and protease. The study of the mechanism of the integrase reaction may also contribute to the further development of gene therapy using lentiviral vectors. The enzymatic activities of HIV-1 integrase will be studied as well as the multimerisation.

Date submitted: 25th August 2002 **Antonie J.W.G. Visser, Ph.D.**

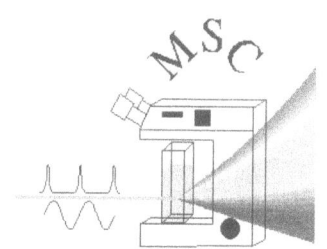

MicroSpectroscopy Centre, Wageningen University,
Dreijenlaan 3, 6703 HA Wageningen, The Netherlands.
And Department of Structural Biology, Vrije Universiteit,
De Boelelaan 1087, 1081 HV Amsterdam, The Netherlands.
Tel: +31 317 482862   Fax: +31 317 484801
Ton.Visser@laser.bc.wau.nl
www.mscwu.nl

Specialty Keywords: Flavoproteins, Fluorescence fluctuations,
Time-resolved fluorescence.

The mission of our MicroSpectroscopy Centre is to strengthen the Dutch infrastructure in optical microspectroscopy, in particular fluorescence. We offer universities, research institutes and industrial companies microspectroscopic state-of-the-art facilities in which biomolecular interactions can be studied such as those among proteins, carbohydrates, lipids, nucleic acids, metabolites, either in isolated form or within cells. Current research is focused on: signal transduction in plants; characterization of plant pathogen resistance genes; gene display technology with high throughput screening; redox biochemistry in complex media and characterization of mesoscopic systems in food sciences.

Structural dynamics of green fluorescent protein alone and fused with a single chain Fv protein. M.A. Hink, R.A. Griep, J.W. Borst, A. van Hoek, M.H.M. Eppink, A. Schots and A.J.W.G. Visser (2000) J. Biol. Chem. 275, 17556-17560.

Date submitted: 23rd August 2002 **Radka S. Vladkova, Ph.D.**

Section Model Membranes, Institute of Biophysics,
Bulgarian Academy of Sciences,
Acad. G. Bonchev Str., Bl. 21, Sofia 1113,
Bulgaria.
Tel: +359 2 979 3694   Fax: +359 2 971 2493
vladkova@obzor.bio21.bas.bg
www.bio21.bas.bg/ibf

Specialty Keywords: Chlorophylls, Membranes, Fluorescent probes.

Intermolecular interactions, organization and dynamics of both the fluorescing molecules (e.g. Chlorophylls, 1,8-ANS) and the medium where they are imbedded (solvents, mixtures, low-temperature matrices, membrane lipid-water structures, and photosynthetic membranes) by using the full arsenal of fluorescence characteristics estimated from steady-state and time-resolved emission spectroscopy, as well as those from hole-burning and site-selection spectroscopy.

R. Vladkova (2000). Chlorophyll *a* self-assembly in polar solvent-water mixtures. *Photochem. Photobiol.*, **71**(1), 71-83.

R. Vladkova, K. Teuchner, D. Leupold, R. Koynova and B. Tenchov (2000). Detection of the metastable rippled gel phase in hydrated phosphatidylcholine by fluorescence spectroscopy. *Biophys. Chem.*, **84**(2), 159-166.

Date submitted: 20th August 2002

Peter Vöhringer, Ph.D.

Max-Planck-Institute for biophysical Chemistry,
Biomolecular and Chemical Dynamics Group,
Am Fassberg 11, Göttingen, 37077,
Germany.
Tel: +49 (551) 201 1333   Fax: +49 (551) 201 1341
pvoehri@gwdg.de
www.mpibpc.gwdg.de/abteilungen/072

Specialty Keywords: Femtosecond spectroscopy, Condensed matter.

Current interests: Dynamics of structural relaxations in biological environments. Ultrafast primary events involved in bioluminescence. Proton, electron, and energy transfer in condensed phase systems. Coherence in liquid phase chemical reactions. Molecular dynamics of liquids.

K. Winkler, J. Lindner, and P. Vöhringer (2002) Low-frequency depolarized Raman-spectral density from femtosecond optical Kerr-effect experiments: Lineshape analysis of restricted translational modes, *Phys. Chem. Chem. Phys.* **4**, 2144-2155.

K. Winkler, J. Lindner, V. Subramaniam, T.M. Jovin, and P. Vöhringer (2002) Ultrafast dynamics in the excited state of green fluorescent protein (wt) studied by frequency-resolved femtosecond pump-probe spectroscopy, *Phys. Chem. Chem. Phys.* **4**, 1072-1081 (2002).

Date submitted: 29th August 2002

Michael Wahl, Ph.D.

PicoQuant GmbH,
Rudower Chaussee 29,
12489 Berlin,
Germany.
Tel: +49 30 6392 6562   Fax: +49 30 6392 6561
wahl@pq.fta-berlin.de

Specialty Keywords: Time-correlated single photon counting, Time-resolved fluorescence, Correlation spectroscopy, Single molecule detection.

M.W. is working as a senior scientist and project leader at PicoQuant GmbH. His research focuses on the development of instrumentation for time-correlated single photon counting. These instruments are applied in ultra-sensitive analysis down to the single molecule level. Recent projects include data acquisition systems for time-resolved fluorescence microscopy and advanced data analysis algorithms for fluorescence correlation spectroscopy and fluorescence lifetime imaging.

Böhmer M.; Wahl M.; Rahn H.J.; Erdmann R.; Enderlein J. "Time-resolved fluorescence correlation spectroscopy" *Chem. Phys. Lett.*, vol 353, 5-6, 439-445 (2002).

Böhmer M.; Pampaloni F.; Wahl M.; Rahn H.J.; Erdmann R.; Enderlein J. "Time-resolved confocal scanning device for ultrasensitive fluorescence detection" Rev. Sci. Instrum. 72, 4145-52 (2001).

Date submitted: 28th August 2002 **Peter Wardman, Ph.D., D.Sc.**

Cancer Research UK *Free Radicals* Research Group,
Gray Cancer Institute, PO Box 100, Mount Vernon Hospital,
Northwood, Middx HA6 2JR,
U.K.
Tel: +44 (0)1923 828611   Fax: +44 (0)1923 835210
wardman@gci.ac.uk
www.gci.ac.uk

Specialty Keywords: Free radicals, Oxidative stress, Radiation chemistry.

My interests focus on the roles of free radicals in cancer biology, particularly the chemistry of cellular oxidative stress and the detection of free radicals or their products in biological systems. Radiation-produced free radicals are of special interest, as are the kinetics of radical reactions. Pulse radiolysis, stopped-flow rapid-mixing and EPR are to characterize reaction kinetics. The chemistry of fluorescent probes that are putative 'reporters' of oxidative and nitrosative stress is of current interest.

Wardman, P., *et al.*, 2002, Pitfalls in the use of common luminescent probes for oxidative and nitrosative stress. *Journal of Fluorescence*, **12,** 65-68.

Ford, E., *et al.*, 2002, Kinetics of the reactions of nitrogen dioxide with glutathione, cysteine, and uric acid at physiological pH. *Free Radical Biology & Medicine*, **32,** 1314-1323.

Date submitted: 4th September 2002 **Watt W. Webb, Sc.D.**

Applied & Engineering Physics, Cornell University,
223 Clark Hall,
Ithaca, NY 14853,
USA.
Tel: 607 255 3331   Fax: 607 255 7658
www2@cornell.edu
www.aep.cornell.edu/drbio

Specialty Keywords: Biophysics, Biomedical, Optics.

The aim of our research is to understand, at the molecular level, the dynamics of basic biophysical processes. The continual challenge is to detect the exquisite subtlety of biomolecular signals and to broaden the paradigms of physical science to encompass biological complexity. The creation of new physical instrumentations addresses this challenge. We study the dynamics of biophysical processes in living cells using modern physical optics such as fluorescence correlation spectroscopy and nonlinear laser scanning microscopy.

Magde, D., Webb, W. W. & Elson, E. Thermodynamic Fluctuations in a Reacting System - Measurement by Fluorescence Correlation Spectroscopy. *Physical Rev Lett* **29**, 705-708 (1972).

Denk, W., Strickler, J. H. & Webb, W. W. Two-Photon Laser Scanning Fluorescence Microscopy. *Science* **248**, 73-76 (1990).

Westman, G.
Widengren, J.

Date submitted: 29th August 2002

Gunnar Westman, Ph.D.

Department of Chemistry and Bioscience,
Chalmers University of Technology,
S-412 96 Gothenburg,
Sweden.
Tel: 46 31 7723072   Fax: 46 31 7723657
westman@oc.chalmers.se
www.oc.chalmers.se

Specialty Keywords: Synthesis, Cyanine dyes,
Benzophenoxazine.

Current interests: Design and synthesis of new fluorescent molecules for the detection and studies of biological systems. Currently we design fluorescent probes that bind in the minor groove of nucleic acids. We also develop fluorescent dyes that show specific staining of cells.

Svanvik, N., Westman, G., Wang, D. and Kubista M. Anal. Biochem. 281, 26-35 (2000) .

Isacsson J and Westman G Tet. Lett. 42, 3207-3210 (2001).

Date submitted: 30th August 2002

Jerker Widengren, Ph.D., M.D.

Dept Medical Biophysics, MBB,
Scheeles v. 2, Karolinska Institutet,
171 77 Stockholm,
Sweden.
Tel: +46 8 7286815   Fax: +46 8 326505
jerker.widengren@mbb.ki.se

Specialty Keywords: FCS, Single Molecule Spectroscopy.

Current research: Development of techniques and applications of Fluorescence Correlation Spectroscopy (FCS) and single molecule Multi-parameter Fluorescence Detection (smMFD). Monitoring and characterization of transient photophysical states, conformations and conformational fluctuations of biomolecules. Detection, characterization and diagnostics of sparse amounts of biomolecules on cell surfaces and in body fluids.

Widengren J, Schweinberger E, Berger S, and Seidel C: J. Phys. Chem., 105, 6851-6866, 2001
Widengren J, Mets, Ü: Conceptual basis of FCS and related techniques as tools in bioscience p. 69-119, in Single Molecule Detection in Solution, Eds. Zander, Enderlein, Keller, Wiley VCH 2002.

Date submitted: 11th September 2002

Gert J. Wilgenhof, Ing.

Varian BV,
Boerhaaveplein 7, Bergen op Zoom,
Postbus 250, 4600 AG,
The Netherlands.
Tel: +31 164282800 Fax: +31 164282828
Gert.Wilgenhof@Varianinc.com
www.varianinc.com

Specialty Keywords: Fluorometer, Spectrofluorometer, Applications.

Varian offers high quality products for measuring fluorescence in many applications. Especially the Cary Eclipse fluorometer offers every wavelength for analyzing fluorescence, phosphorescence and chemi-luminescence with excitation and emission scans as well as 2D / 3D plots. Temperature control, polarization, fiber optic and wellplate options are available. With the instrument knowledge Varian participates in research projects and helps with developing new applications. The Varian office in Bergen op Zoom is equipped with all the necessary tools to make your fluorescent application work.

Please contact: Gert Wilgenhof – Cary Eclipse specialist The Netherlands.

Date submitted: 6th August 2002

Stuart A. Windsor, Ph.D.

Biotechnology Team, National Physical Laboratory,
Queens Road, Teddington,
Middlesex, TW11 0LW,
UK.
Tel: +44 (0)20 8943 7085 Fax: +44 (0)20 8614 0531
Stuart.Windsor@npl.co.uk
www.npl.co.uk/biotech

Specialty Keywords: Fluorescence standards, Quantum dots, Biological fluorescence, FCS, Multparameter fluorescence.

Current Research Interests: My research is focused on the validation and standardization of fluorescence based techniques (particularly those used in the bioscience), and the development of new biological characterization methods based on fluorescence measurement. Current active research includes the development of quantum dot fluorescence standards; the use of multi-parameter fluorescence measurements for biopharmaceutical characterization; the validation of high-throughput fluorescence measurement methods; and the development of single molecule structural characterization methods based on fluorescence correlation spectroscopy. I work closely with industry and academia, and have recently initiated a major consortium (BEACON) to improve biopharmaceutical characterization methods (incl. fluorescence) used in regulation

Date submitted: 7th March 2002

Otto S. Wolfbeis, Ph.D.

Professor of Analytical & Interface Chemistry,
University of Regensburg,
Institute of Analytical Chemistry, Chemo - and Biosensors,
93040 Regensburg,
Germany.

Specialty Keywords: Chemical sensors, Biosensors, Fluorescentprobes, Interface chemistry, Bioassays, Nanoparticles.

Current Research: (Fiber) optic chemical sensors and planar sensors for blood gases and blood electrolytes, enzyme based biosensors for glucose, lactate and urea; fluorescent probes for immunoassay via FRET and PRET; decay time based biosassays; beads as labels for proteins and polynucleotides; multiplexing of bead arrays via lifetime and color (arrays); self-assembled monolayers on gold films; biosensor arrays using gold films; molecular imprints and footprints.

New Type of Phosphorescent Nanospheres for Use in Advanced Time-Resolved Multiplexed Bioassays, J. M. Kuerner, I. Klimant, Ch. Krause, E. Pringsheim & O. S. Wolfbeis, *Anal. Biochem.* **297** (2001) 32-41.

.Novel Diode Laser-Compatible Fluorophores and Their Application to Single Molecule Detection, Protein Labeling and Fluorescence Resonance Energy Transfer Immunoassay", B. Oswald, M. Gruber & O. S. Wolfbeis, *Photochem. Photobiol.* **74** (2001) 237-242.

Date submitted: 16th September 2002

Danuta Wróbel, Ph.D.

Institute of Physics, Poznan University of Technology,
Nieszawska 13A, Poznan,
Poland, 60-965,
Poland.
Tel: +48 61 669 3179 Fax: +48 61 669 3201
wrobel@phys.put.poznan.pl
www.put.poznan.pl

Specialty Keywords: Molecular Physics, Molecular Spectroscopy, Organic dyes.

The study of spectral properties of synthetic organic dyes and chlorophyll pigments in isotropic and anisotropic media to follow: mechanisms of radiative and non-deactivation processes of porphyrins and phthalocyanines, of porphyrin-melanin systems, mechanisms of generation of the photovoltaic effects in photoelectrochemical cells based on synthetic organic dyes and biological materials, Langmuir-Blodgett layers, optical and IR studies, organic photovoltaics, photodynamic therapy.

D.Wróbel, *et al.*, Fluorescence and time-resolved delayed luminescence of porphyrins in organic solvents and polymer matrices, J.Fluorescence, 8 (1998) 191-198.

A.Boguta, D.Wróbel, Fluorescein and phenolophthalein-Correlation of fluorescence and photovoltaic properties, J. Fluorescence, 11 (2001), 131-139.

Date submitted: 24th August 2002

Li Yao-Qun, Ph.D.

Department of Chemistry,
Xiamen University,
Xiamen 361005,
China.
Tel & Fax: +86 592 2185875
yqlig@xmu.edu.cn

Specialty Keywords: Fluorescence, Multi-component analysis.
The research fields include molecular fluorescence spectroscopy and its application in environmental and biological analysis, multi-component analysis, and surface analysis. Special interests have focused on the development, instrumentation and application of some fluorescence techniques, such as synchronous fluorescence spectroscopy, multi-dimensional fluorescence, derivative technique, reflection fluorescence and confocal microscopy.

Derivative matrix isopotential synchronous fluorescence spectroscopy for the direct determination of 1-hydroxypyrene as a urinary biomarker of exposure to polycyclic aromatic hydrocarbon, with Wei Sui, Chun Wu, Li-Jun Yu, *Anal Sci.*, **2001**, 17(*1*),167.

Spectral fluctuation and heterogeneous distribution of porphine on the water surface, with Maxim. N. Slyadnev, Takanori Inoue, Akira Harata and Teiichiro Ogawa, *Langmuir*,**1999**, 15(*9*), 3035.

Date submitted: 29th August 2002

Sergiy M. Yarmoluk, Ph.D.

Institute of Molecular Biology and Genetics of NAS of Ukraine, Zabolotnogo Str. 150,
Kyiv, 03143,
Ukraine.
Tel: (+38 044) 252 23 89   Fax: (+38 044) 252 24 58
sergiy@yarmoluk.org.ua
www.yarmoluk.org.ua

Specialty Keywords: Fluorescent probes, Cyanine dyes, Nucleic acids.
The research interests of Dr. Yarmoluk are connected with fluorescent detection of biological molecules. In the Nucleic Acids Chemistry Lab guided by Dr. Yarmoluk the series of cyanine dyes as novel fluorescent probes for nucleic acids and proteins detection were proposed, and novel methods for biomolecules labeling with cyanine dyes were developed [1]. The interaction of the cyanine dyes with nucleic acids are studied [2].

S.M. Yarmoluk, A.M. Kostenko, I.Ya. Dubey (2000) *Bioorg. Med. Chem. Lett.* 10, 2201-2204.
S.M. Yarmoluk, S.S. Lukashov, T.Yu. Ogul'chansky, M.Yu. Losytskyy, O.S. Kornyushyna (2001) *Biopolymers* 62, 219-227.

Date submitted: 22nd August 2002

Liming Ying, Ph.D.

Department of Chemistry,
University of Cambridge,
Lensfield Road, Cambridge,
CB2 1EW, UK.
ly206@cam.ac.uk
www-klenerman.ch.cam.ac.uk

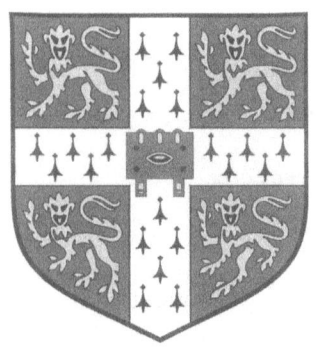

Specialty Keywords: Single Molecule Fluorescence,
Fluorescence Resonance Energy Transfer (FRET).

My research focuses on applying single molecule fluorescence spectroscopy, especially single molecule FRET and FRET correlation spectroscopy to study structural heterogeneity and conformational dynamics of biomolecules including DNA hairpins and G-quadruplexes. I am also developing novel fluorescence methods such as single molecule fluorescence coincidence to detect the interaction of proteins with nucleic acids. My current projects aim to tackle the structure and mechanism of DNA polymerase and telomerase at the single molecule level.

Wallace M. I., Ying L. M., Balasubramanian S., Klenerman D. Non-Arrhenius Kinetics for the Loop Closure of a DNA Hairpin, *Proc. Natl. Acad. Sci. USA,* **98**, 5584 (2001).

Ying L. M., Xie X. S., Fluorescence Spectroscopy, Exciton Dynamics, and Photochemistry of Single Allophycocyanin Trimers, *J. Phys. Chem. B* **102**, 10399 (1998).

Date submitted: 31st May 2002

Stephen Yue, Ph.D.

Organic Chemistry, Molecular Probes, Inc.,
4849 Pitchford Avenue, Eugene,
Oregon, 97402,
USA.
Tel: (541) 465 8300   Fax: (541) 344 6504
stephen.yue@probes.com
www.probes.com

Specialty Keywords: Fluorescence.

Earned Ph.D. in Organic Chemistry from Oregon State University (1982). Principal Scientist at Molecular Probes, Inc. and Inventor of SYBR series nucleic acid stains and some NIR Alexa Fluor dyes. Other activities are in development of new fluorescent dyes.

Date submitted: 3rd September 2002

Christoph C. Z. Zander, Ph.D.

AG Biochemistry, University of Siegen,
Adolf-Reichwein-Strasse 2, Siegen,
D-57068,
Germany.
Tel: +49 271 740 4129
zander@chemie.uni-siegen.de
www.uni-siegen.de
Specialty Keywords: Laser cooling, Single molecule detection, Anti Stokes fluorescence.

My group is working since 1991 in the field of fluorescence. The mayor topics of this works are laser cooling by anti Stokes fluorescence (see ref. 1) and single molecule detection (see ref. 1).

Cooling of a Dye Solution by Anti-Stokes Fluorescence, C. Zander, K.H. Drexhage, Advances in Photochemistry, Volume **20**, John Wiley & Sons (1995) 59.

Single Molecule-Detection in Solution: Methods and Applications, eds. Ch. Zander. J. Enderlein, R.A. Keller, Wiley-VCH Verlag Berlin GmbH, S. 247 – 272, Berlin 2002.

Date submitted: 30th August 2002

Jie Zheng, Ph.D.

Howard Hughes Medical Institute,
Department of Physiology and Biophysics,
University of Washington, Box 357290,
Seattle, WA 98195-7290, USA.
Tel: 206 543 8076   Fax: 206 543 0934
jzheng@u.washington.edu

Specialty Keywords: Patch-clamp fluorometry, Ion channel, Membrane protein, FRET.

My research has been focused on applying fluorescence techniques to the study of membrane protein dynamics. Through a combination of classic patch-clamp current recordings and site-specific fluorescence recordings, the conformational rearrangements of ion channels are monitored and related to the functional states of these membrane proteins. Currently I am using FRET to study gating and modulation of the cyclic nucleotide-gated channels that mediate sensory transduction of both visual and olfactory systems.

J. Zheng, and W.N. Zagotta (2000) Gating rearrangements in cyclic nucleotide-gated channels revealed by patch-clamp fluorometry, *Neuron*, **28**, 369-374.

Zilles, A.
Zozulya, V. N.

Date submitted: 4th September 2002

Alexander Zilles, Ph.D.

ATTO-TEC GmbH,
57008 Siegen,
Germany.
Tel: +49 (0) 271 740 4735
zilles@atto-tec.de
www.atto-tec.com

Specialty Keywords: Fluorescence, Fluorescent organic dyes, Biolabeling, Bioanalytics, Photofading, UV-absorber.

My research interests are based on the design and synthesis of novel fluorescent dyes. Individual functionalisation of these dyes make them highly suitable for bioanalytical applications e.g. biolabeling of nucleotides, proteins, etc.
I am further interested in the development and investigation of detergent additives to prevent photodegradation of dyed fabrics in particular cellulosic based fibers e.g. cotton.

J. Arden-Jacob, J. Frantzeskos, N. U. Kemnitzer, A. Zilles, K. H. Drexhage, *Spectrochim. Acta,* **57A**, 2271-2283 (2001). A. Zilles, PhD Theses *"The design and synthesis of detergant additives for the photo-chemical protection of dyed fabrics."* University of Leeds, Department of Colour Chemistry (2002).

Date submitted: 30th August 2002

Victor N. Zozulya, Ph.D.

Department of Molecular Biophysics,
B.Verkin Institute for Low Temperature Physics and Engineering, NAS of Ukraine,
47 Lenin Ave., Kharkov, 61103, Ukraine.
Tel: 380 572 308 534 Fax: 380 572 322 370
zozulya@ilt.kharkov.ua

Specialty Keywords: Fluorometry, Dye-Nucleic acid complexes, Dye-oligonucleotide conjugates.

Investigation of fluorescent and binding properties of intercalating dyes and drugs in complexes with polynucleotides and nucleic acids. Utilization of covalently attached dyes as fluorescent probes and stabilizers of antisense and antigene oligonucleotide hybridization.

Victor Zozulya (1999) Fluorescence properties of intercalating neutral chromophores in complexes with polynucleotides of various base compositions and secondary structures. *J. Fluorescence* 9, 363 - 366.
V. Zozulya, A. Shcherbakova and I. Dubey (2000) Calculating helix-to-coil transitions of duplexes formed by phenazine-conjugated oligonucleotide, using fluorescence melting data. *J. Fluorescence* 10, 49 - 53.

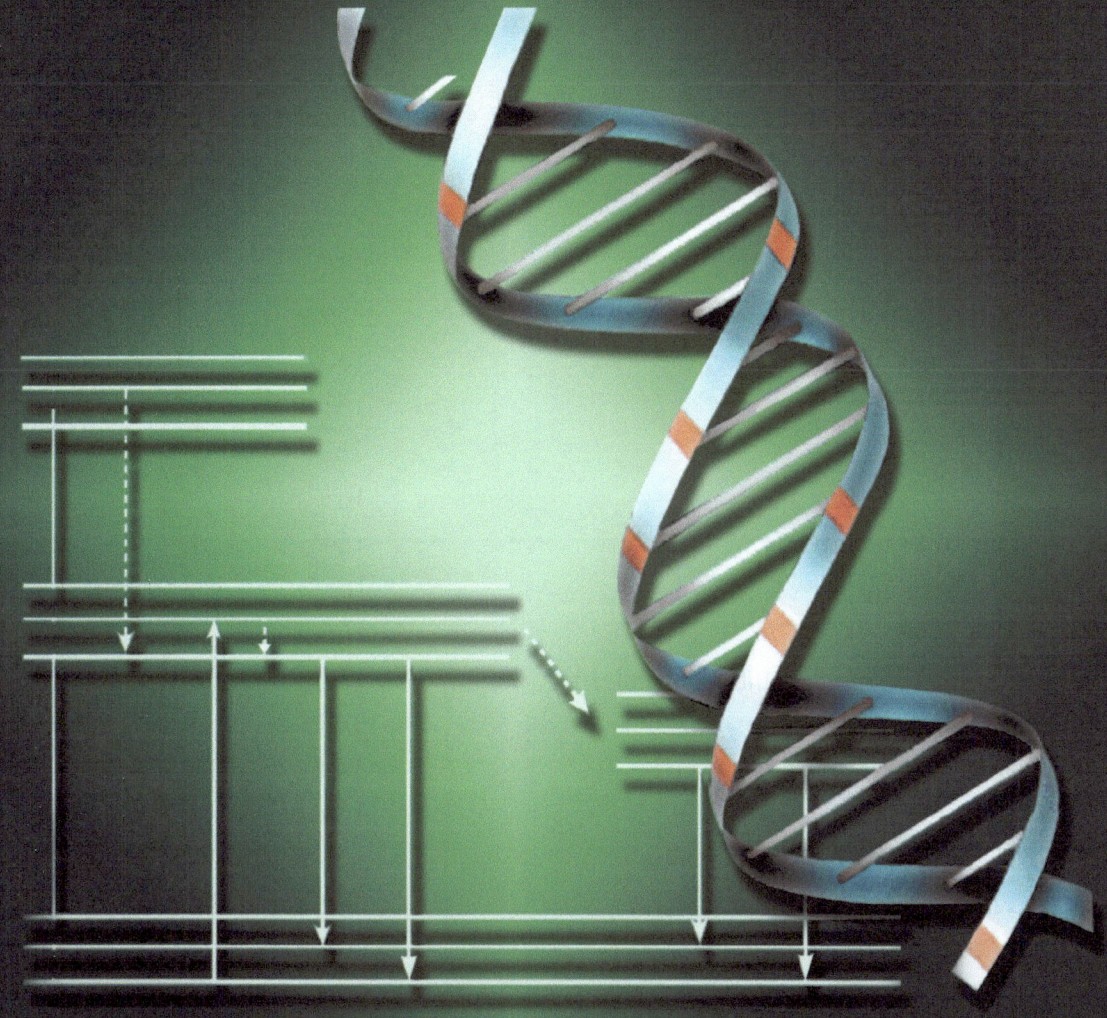

2003

Who's Who in
Fluorescence

Company Entries

Date submitted: 11th July 2002

ATTO-TEC GmbH

Fluorescent Dyes & BioLabeling,
Schanzenweg 50, D-57076 Siegen,
Germany.
Tel: +49 (0)271 740-4995, Fax: +49 (0)271 740-4994
info@atto-tec.com
www.atto-tec.com

Specialty Keywords: Fluorescent Dyes, DNA, Labeling Services

Fluorescent Dyes & BioLabeling

We are specialists in the development of fluorescent dyes and labeling systems. We also prepare DNA and protein labels based on fluorescent dyes. Our services are fully customized to best meet your individual needs.

We are developing and producing laboratory equipment and an on-chip multianalysis sensor for detecting antigens, e.g. pathogens, in liquids. In addition we work on single-molecule detection through highly sensitive molecular-biological detection systems.

Our company works for the biotechnology sector in the pharmaceutical industry and universities.

Products

- fluorescent dyes and labeled oligonucleotides

- intelligent DNA-probes: GenePin

Immun-o-mat

- fully automated immuno assays on a biochip platform

LightStation

- sophisticated microscope for single molecule detection.

ATTO-TEC

BIOLABELING AND
ULTRAANALYTICS

Date submitted: 5th September 2002

Avanti Polar Lipids, Inc.

700 Industrial Park Drive
Alabama, 35007-9105
USA
Tel: (800) 227 0651, (205) 663 2494 Fax: (800) 229 1004, (205) 663 0756
Email: info@avantilipids.com URL: www.avantilipids.com

Specialty Keywords: Lipid, Phospholipid, Sphingolipid.

For over 30 years, Avanti Polar Lipids, Inc. has been supplying lipids to researchers and pharmaceutical companies around the world. The range of products includes over 1,000 lipids, all made with Avanti's famous purity. Research clients are encouraged to contact us about custom synthesis of products not listed in our catalog; we are pleased to offer our Pharmaceutical clients products which follow cGMP guidelines.

Avanti's selection of chemicals and range of services include -

2-DIMENSIONAL CRYSTALLOGRAPHY OF PROTEINS.
ANALYTICAL SERVICES.
BICELLE FORMING LIPIDS.
¹³C LABELED PC.
CERAMIDES.
CERAMIDE DERIVATIVES – PHOSPHORYLATED.
CONJUGATION OF PROTEINS, PEPTIDES & DRUGS TO LIPOSOMES.
DEUTERIUM LABELED LIPIDS.
DIACYLGLYCERIDE PYRO PHOSPHATES (DGPP).
EXTRUSION EQUIPMENT FOR LIPOSOME PREPARATION.
FLUORESCENT PHOSPHOLIPIDS.
FLUORESCENT CHOLESTEROL.
GANGLIOSIDES.
GLYCOLIPIDS.
LIPID MIXTURES (FOR CELL CULTURE & CLINICAL APPLICATIONS).
LIPID A (HIGH PURITY).
LIPOTRANSFER.
MALEIMIDE-CONTAINING LIPIDS.
NICKEL CHELATING LIPIDS FOR BINDING HIS-TAGGED PROTEINS.
OXY-STEROLS.
PHOSPHATIDYLINOSITOLS.
PHOSPHOLIPIDS & SPHINGOLIPIDS.
PIP & PIP2.
SECONDARY STRUCTURAL ANALYSIS OF PROTEINS & MACROMOLECULES.

ADJUVANT FOR VACCINES.
ANTIGENIC LIPIDS.
BIOTINYLATED LIPIDS.
CATIONIC REAGENTS.
CERAMIDE DERIVATIVES – GLYCOSYLATED.
CEREBROSIDES.
TRANSFECTION OF GENETIC MATERIAL.
SYNTHETIC LIPIDS.
DIACYL GLYCEROLS.
DRUG DELIVERY VEHICLES.
SPHINGOSINE DERIVATIVES - PHOSPHORYLATED.
FLUORESCENT SPHINGOLIPIDS.
FUNCTIONALIZED PEG-LIPIDS.
GLYCEROL BASED LIPIDS.
LANTHANIDE CHELATING LIPIDS.
SPHINGOSINE.
SECOND MESSENGERS.
LIPOSOME.
LYSOPHOSPHATIDIC ACID.
NATURAL LIPIDS.
SPHINGOSINE DERIVATIVES-GLYCOSYLATED POLYMERIZABLE LIPIDS.
PEG-LIPIDS.
PHOSPHATIDYLSERINE.
PHYTOSPHINGOSINE.
SPHINGOMYELIN.

Date submitted: 4th September 2002

Berry & Associates, Inc.

2434 Bishop Circle East
Dexter, Michigan 48130
U.S.A.
(734) 426-3787 or (800) 357-1145 (Toll-free in US)
alchem@berryassoc.com
www.berryassoc.com

Specialty Keywords: Fluorescent Dyes, Carboxyfluoresceins, Carboxytetramethylrhodamines

---

PROVIDING 5-, 6- ISOMERICALLY PURE DYES FOR YOUR LABELING NEEDS
(visit www.berryassoc.com for competitive pricing)

Berry & Associates, Inc. is a leading source of nucleosides, modified nucleosides and related compounds. We also offer a number of high isomerically pure dyes and their N-Hydroxysuccinimide derivatives (see examples at left) for researchers to incorporate into oligonucleotides or proteins.

ALSO FEATURING

- Biotins
- Carbohydrates
- Fluoresceins
- Heterocycles
- Linkers
- Psoralens
- Purine Nucleosides
- Purine Nucleoside Analogs
- Pyrimidine Nucleosides
- Pyrimidine Nucleoside Analogs
- Rhodamines

BERRY&ASSOCIATES
Synthetic Medicinal Chemistry

Date submitted: 12th September 2002

Carl Zeiss MicroImaging, Inc.
One Zeiss Drive
Thornwood, NY 10594
800-233-2343; fax: 914-681-7446
micro@zeiss.com
www.zeiss.com/micro

Specialty Keywords: Fluorescence; Confocal; FRET; 6D; GFP.

The field of fluorescence microscopy in biological research has seen an unprecedented evolution over the past 10 years. As a result, we are able to answer questions that we never dreamed possible. But the resulting demands upon the fluorescence microscope have also evolved in an unprecedented way. No longer do we ask yes or no. Today we ask how much, how many, how fast. We are observing GFP in whole animals (Zeiss M2Bio stereomicroscope) as well as single molecule fluorescence (Zeiss Axiovert 200 inverted microscope or FCS). We can scan tissue sections or gene microarrays. We no longer look at samples in 2D. Today, we demand 6D imaging.

But regardless of the range of the extremes that we are working in, the fundamental demands upon the microscope remain common:

1. Outstanding Signal to Noise, a result of the symphony of:
      a. Objective correction
      b. Fluorescence filter selectivity
      c. Efficiency of illumination light train
      d. Efficiency of imaging light train
2. Ergonomy of Platform Design
3. Flexibility of Platform
4. 3D Imaging Abilities

Whether Macro or Micro; High Speed or High Resolution; Wide Field, Deconvolution or Confocal Imaging. You will need a partner who understands the emerging challenges of today and tomorrow. **Carl Zeiss: Your Partner in Fluorescence Microscopy.**

Date submitted: 12th September 2002

Division of Chemodex Ltd.

Waldkircherstr. 7   Tel.: +41 71 244 48 25
9213 Hauptwil   Fax +41 71 244 48 26
Switzerland   info@fluoroprobes.com
www.fluoroprobes.com

Specialty Keywords:
Fluorescent Probes & Labels, Synthesis, Switzerland

Need new markers, probes, stains, labels? Looking for a cheap source of common fluorescent chemicals worldwide? We supply analytical reagents used mainly in fluorescence applications.

Our strength is

Synthesis and sourcing of laboratory chemicals (in particular fluorescent markers, probes, labels and stains).

All substances distributed by Fluoroprobes are prepared under a strict quality assurance in our own laboratories in Europe. In addition we co-operate with external laboratories.

Our product range:

- new items (not yet published in the literature)
- published compounds known from the literature (but not yet commercially available)
- established and commercially available chemicals

The chemicals are either

- available from stock in Switzerland
- made by custom synthesis (according to the customer order and specifications)
- sourced worldwide for you

Due to our low overhead, we can offer competitive prices!
Take advantage from our low bulk prices!
We already supply resellers in bulk quantities!

Date submitted: 1st June 2002

ISS

1602 Newton Drive, Champaign,
IL, 61822, USA.
Tel: 217 359 8681 Fax: 217 359 7879
iss@iss.com
www.iss.com

Specialty Keywords: Steady-state and Lifetime Spectrofluorometers, Fluorescence Sensing, Lifetime Imaging, Fluorescence Microscopy, Fluorescence Correlation Spectrometers, Near-Infrared Oximetry

ISS was founded in 1984. It is a leader in the development of highly sensitive fluorescence instrumentation for research, clinical, and industrial applications. Recent product innovations at ISS include:

- ALBA – Single and Dual Channel Fluorescence Correlation Spectroscopy System
- Chronos – Lifetime spectrometer utilizing laser diodes or light emitting diodes as light sources

Along with the fully automated K2 Multifrequency Phase and Modulation fluorometer, and the PC1 Photon Counting spectrofluorometer, ISS offers over 30 accessories and modular components for spectrofluorometers, the OxiplexTS Tissue Spectrometer, and Imagent, for functional brain mapping.

The company's current customer base includes universities and industry in the Americas, Europe, Asia and Australia. The ISS facilities are located in Champaign, Illinois at the Interstate Research Park. The 22,000 square foot facility houses the offices, the mechanical manufacturing workshop, an electronics and optics facility, a product-testing laboratory, an engineering area and a visitor training center.
To find out more about ISS, please visit our website at www.iss.com

ISS Achievements

The following table lists the instrumentation introduced by ISS since 1985.

1985	• First commercial Multifrequency Phase Fluorometer (MPF), the GREG200™
1987	• Installation of the first MPF using the harmonic content of mode-locked lasers
1989	• First fully automated Multifrequency Phase Fluorometer, the K2™ MPF
1991	• First MPF with digital filter for the isolation of the cross-correlation frequency and user-selectable cross-correlation frequency
	• Introduction of K2-FastScan™ capable of acquiring time-resolved fluorescence data in a minute using either a lamp or a laser as an excitation light source.
1992	• First MPF equipped with diode array detector for simultaneous multifrequency-multiwavelength fluorescence data acquisition.
1997	• Introduction of Model 96208 Two-channel Oximeter
1999	• Introduction of ALBA, a Fluorescence Correlation Spectrometer
2001	• Introduction of CHRONOS a lifetime spectrometer based on laser diodes and LEDs
	• Introduction of Galaxy, a microscope based spectrofluorometer capable of steady-state fluorescence measurements, lifetime measurements and FCS simultaneously.
	• Introduction of Imagent, functional brain imaging system

Innovations in Fluorescence

Date submitted: 9th September 2002

163

Date submitted: August 29th 2002

Lambert Instruments

Turfweg 4
9313 TH, Leutingewolde
The Netherlands
Tel: +31-(0)50-5018461 Fax: -5010043
E-mail: info@lambert-instruments.com
www.lambert-instruments.com

Specialty Keywords: Fluorescence Lifetime Imaging Microscopy - FLIM, Fluorescence Resonance Energy Transfer - FRET, frequency domain, LED.

Lambert Instruments specializes in low light level image detectors and systems for scientific applications making use of image intensifiers, standard and custom made.

Lambert Instruments' **LIFA Fluorescence Lifetime Imaging Attachment** is a system that can be attached to any wide field fluorescence microscope, allowing fluorescence image aquisition and the generation of lifetime images. In contrary to other systems, working in the time domain, the LIFA system works in the frequency domain, having the advantage of a more efficient use of the available photons.

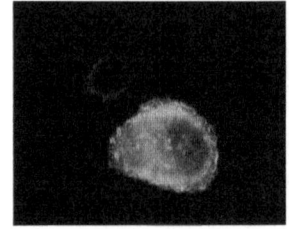

MCF7 cells with
ErbB.1-GFP as donor
and Py72/Cy3 as
acceptor

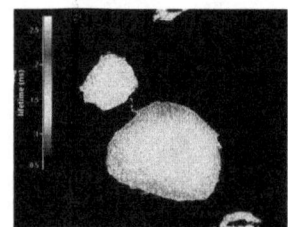

Intensity

Lifetime

Why lifetime imaging?

Quantitative fluorescence microscopy uses the intensity of the fluorescence to extract information about the local concentrations of molecules that are labeled with fluorescent probes. This technique suffers from the fact that the fluorescence of the probe is permanently destroyed by light-induced conversion of the probe material to a non-fluorescent compound. This photochemical process is called "bleaching" and makes it necessary to regulate the excitation dose in an economical way. Another physical property of fluorescent molecules is the fluorescence lifetime (the decay time of the emission after the excitation has been stopped) which depends on the local concentrations of certain molecules or ions. Changes in fluorescence efficiency as caused by bleaching are not accompanied by changes in fluorescence lifetime. Fluorescence lifetime imaging microscopy (FLIM) merges the information of the spatial distribution of the probe with the lifetime to increase the reliability of the concentration measurements. Additionally, FLIM enables the discrimination of fluorescence coming from different dyes, including auto-fluorescent materials, that exhibit similar absorption and emission properties but showing a difference in fluorescence lifetime. Another application is the study of macro molecular interactions using GFP labeling and the combination of fluorescence resonance engergy transfer (FRET) and FLIM.

Date submitted: 6th September 2002

LaVision
WE COUNT ON PHOTONS

LaVision GmbH
ANNA-VANDENHOECK-RING 19
D37081 GOETTINGEN
GERMANY
TEL. +49 551 900 40 / FAX +49 551 900 4100
INFO@LAVISION.DE
WWW.LAVISION.DE

LaVision Inc.
301 W. MICHIGAN AVE., SUITE 403
YPSILANTI, MI 48179
USA
TEL. 734 485 0913 / FAX 240 465 4306
SALES@LAVISIONINC.COM
WWW.LAVISIONINC.COM

CCD & ICCD Cameras, Imaging Spectroscopy, FLIM

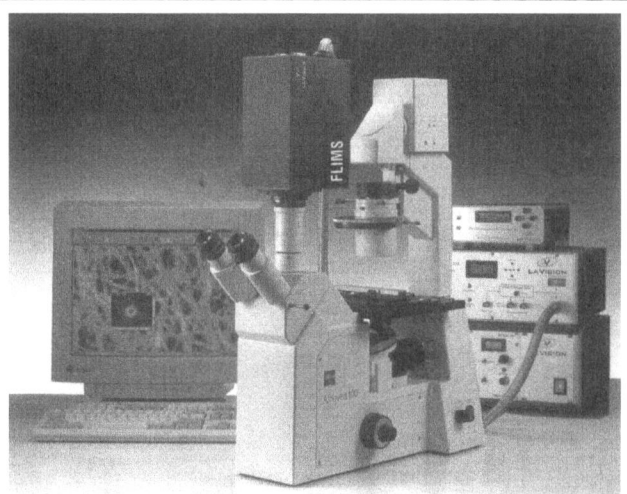

Ultrafast Gated Cameras

PicoStar HR	200 ps @ 110MHz
PicoStar UF	50 ps @ 10kHz
ModStar	Gain Modulation up to 1GHz

▸ Single Photon Sensitivity
▸ Spectral Range 200-900nm
▸ Digitization 12-16 bit
▸ Up to 30 images/s

▸ High performance software for device control, camera functions, system integration, image acquisition, processing and analysis

Integrated Turn-Key Systems For

▸ Time-Resolved Imaging, Spectroscopy and Microscopy
▸ Fluorescence Lifetime Imaging Microscopy (FLIM)
▸ Multifocal Multiphoton Microscopy
▸ Imaging Through Scattering Media
▸ Gating and Ranging (LIDAR, Underwater Imaging)

Date submitted: 28th March 2002

Olis, Inc.

130 Conway Drive, Suites A & B130 Conway Drive,
Bogart, Georgia 30622,
USA,

Tel: 706 353-6547 Fax: 706 353-1972
olis@negia.net
www.olisweb.com

Specialty Keywords: Fluorescence, Absorbance, CD Spectroscopy

---

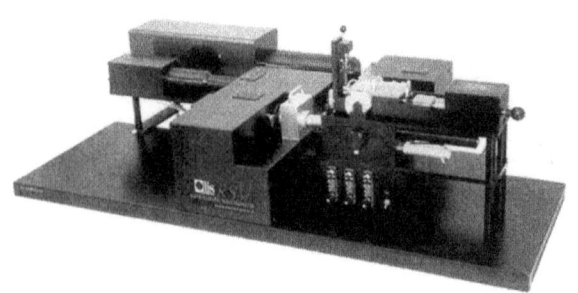

**RSM 1000 Spectrofluorimeter
with Stopped-Flow**

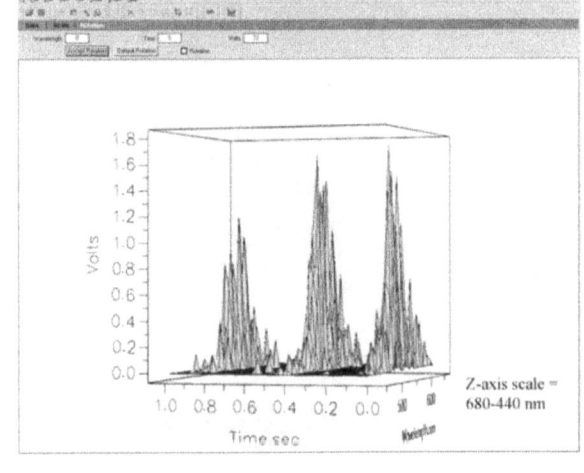

Olis DM 45 Spectrofluorimeter

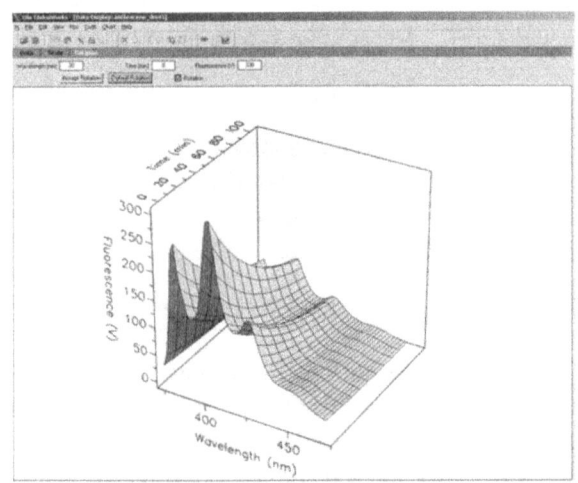

Date submitted: 13th September 2002

Photon Technology International, Inc.

1009 Lenox Drive, Suite 104, Lawrenceville,,
NJ 08648
USA.
Tel: (609) 896 0310
marketing@pti-nj.com
www.pti-nj.com

Specialty Keywords: **Fluorescence, Lifetimes, Ratio, Imaging.**

PTI Products...

Photon Technology International offers complete systems for the three primary areas of fluorescence measurements: **steady state**, **lifetime** and **microscopy/imaging**. Steady state and ratiometric measurements are represented by the **QuantaMaster** line – the world's most sensitive fluorometers. Fluorescence lifetimes are addressed by the **TimeMaster** line, the extremely versatile and powerful time-domain based lifetime fluorometers based on *PTI*'s proprietary technology. And finally the **MicroMaster** offers systems for conventional fluorescence microscopy, fluorescence imaging, as well as specialized microscope-based systems for the measurement of fluorescence lifetimes. Thanks to *PTI*'s **Open Architecture Design**, all of our fluorescence systems are compatible with one another. A QuantaMaster steady state fluorometer purchased today can be easily enhanced with TimeMaster lifetime system capabilities tomorrow. The ability to make measurements with microscopes can be added to a cuvette-based system and vice versa. *PTI* also offers an extensive line of **Optical Building Blocks (OBB)**, which include several types of quarter meter monochromators, cost-effective low-light level intensified CCD cameras, dual digital/analog compact detection systems, various microscope accessories including single and dual channel photometers, nitrogen lasers, nitrogen-pumped dye lasers and frequency doublers and various types of pulsed and continuous light sources.

About PTI...

Photon Technology International Inc. is a public corporation that was established in 1983 to develop light-based instrumentation for fluorescence and phosphorescence spectroscopy and has been instrumental in pioneering many new innovations in the field. *PTI* develops and manufactures its own equipment. The Company, in conjunction with related companies (*PhotoMed GmbH*, *PTI Canada* and *PTI UK*), maintains offices, customer support and service in the U.S., Canada, Germany, Denmark and England. The remainder of our worldwide distribution is handled through company-trained representatives.

Date submitted: 24th June 2002

PicoQuant GmbH

Rudower Chaussee 29,
Berlin,
12489,
Germany.
Tel: +49-(0)30-6392-6560 Fax: -6561
photonics@pq.fta-berlin.de
www.picoquant.com

Specialty Keywords: Pulsed Laser Systems, Photon Counting Equipment, Time-resolved Fluorescence Systems, Single Molecule Detection Microscopes

PicoQuant GmbH is a research and development company based in Berlin-Adlershof, Germany. The company is leading in the field of Single Photon Counting Applications. The product line includes a successful range of fluorescence lifetime systems, photon counting instrumentation and pulsed light sources. PicoQuant developed a compact and easy to use picosecond diode laser with wavelength from the Violet to Infrared, the PicoQuant PDL 800-B. It is unique in power, pulse width and repetition rate and without competition world wide. In order to complement these laser products in TCSPC applications, PicoQuant also developed new technology for time-correlated photon counting, the TimeHarp TCSPC systems on a single PC board. Combining these technologies, the outstanding range of the time resolved-spectrometers was born. A variety from compact systems to high end modular lifetime spectrometers and finally complete epi-fluorescence time-resolved microscopes are available.

Major Products:

- FluoTime 100 - The Smallest Fluorescence Lifetime Spectrometer

- FluoTime 200 - The Modular High-end Fluorescence Lifetime Spectrometer

- MicroTime 100 - Time-resolved Laser Fluorescence Microscope (upright)

- MicroTime 200 - Time-resolved Laser Fluorescence Microscope (inverse and confocal)

- TimeHarp 200 - Single PCI-board for TCSPC

- LDH440, The First True Blue Diode Laser @ 40 MHz

- PDL800-B, Sepia, PLS, MDL 300, FSL 500 - Pulsed Light Sources

Services:

Annual International Workshop on Single Molecule Detection and Ultrasensitive Analysis in

Date submitted: 13th August 2002

Polysciences, Inc.

400 Valley Road
Warrington, PA 18976 USA
Phone: 1-215-343-6484
Fax: 1-215-343-0214
Email: info@polysciences.com
www.polysciences.com

Fluorescent Microparticles, Fluorescent Monomers, Fluorescent Dyes

Polysciences, Inc. manufactures and supplies a full line of fluorescent microparticles, conjugated antibodies, fluorescent dyes, and fluorescent monomers. These products can be used in a variety of applications including flow cytometry, fluorescence microscopy, confocal microscopy, biological assays, and the synthesis of fluorescent polymers.

Polysciences, Inc. products include:

Flow Check® Flow Cytometry Calibration Microspheres
Fluoresbrite® Plain and Functionalized Microspheres
Fluoresbrite® Multifluorescent and Polychromatic Red Microspheres
Fluoresbrite® PolyFluor® and Chemiluminescent Microspheres
Fluoresbrite® Size Range and Color Range Kits

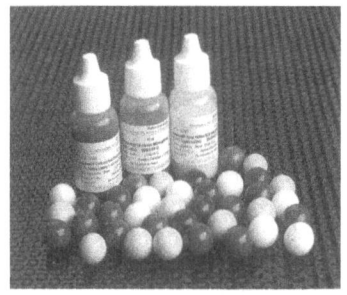

A Selection of Over 60 Different Fluorescent Dyes and Visible Stains

PolyFluor® Fluorescent Monomers:	Ex. Max.	Em. Max.
PolyFluor® 345 (2-Naphthyl methacrylate)	285nm	345nm
PolyFluor® 394 (1-Pyrenylmethyl methacrylate)	339nm	394nm
PolyFluor® 407 (9-Anthracenylmethyl methacrylate)	362nm	407nm
PolyFluor® 497 (O-Methacryloyl Hoechst 33258)	355nm	497nm
PolyFluor® 511 (Fluorescein dimethacrylate)	470nm	511nm
PolyFluor® 512 (3,8-Dimethacryloyl ethidium bromide)	439nm	512nm
PolyFluor® 570 (Methacryloxyethyl thiocarbamoyl rhodamine B)	548nm	570nm

Fluorescent Dyes Conjugated to Streptavidin and Secondary Antibodies

For a full listing of Polysciences, Inc. products request a catalog at www.polysciences.com.

 ULTRA Evolution  TECAN.

Tecan's most advanced multi-functional microplate detection system offers the best levels of detection performance without compromising flexibility.

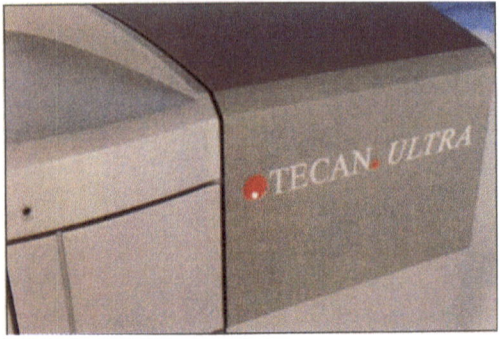

A multifunctional instrument for:
- Fluorescence Lifetime Measurements
- Fluorescence Intensity (UV and VIS range)
- Time "gated" fluorescence
- Homogeneous time resolved fluorescence
- Time resolved FRET
- Fluorescence polarization
- Glow type luminescence
- Absorbance (UV and VIS range)

Introduction

The Ultra Evolution is Tecan's latest development on the Ultra instrument platform. New features are available as user definable options which adds to the flexibility of the existing instrument along with possible upgrade paths should your needs change in the future.

New instrument Highlights

- Fluorescence Lifetime Measurement in Microplates

The already successful Ultra now offers a new technique – fluorescence lifetime measurement in the nanosecond time domain. Tecan have added state-of-the-art laser technology to the optical pathways, providing the user with a unique instrument for assay development.

The laser excitation sources are coupled into the existing optical system, so that performance is still guaranteed with all the existing measurement techniques. As usual with Tecan ULTRA, there are several features designed to optimise the measurement parameters automatically in this new detection mode.

Tecan are able to offer a choice of two laser excitation wavelengths

Unique laser attenuation solution

Z axis focusing

Automatic decay curve fitting algorithm

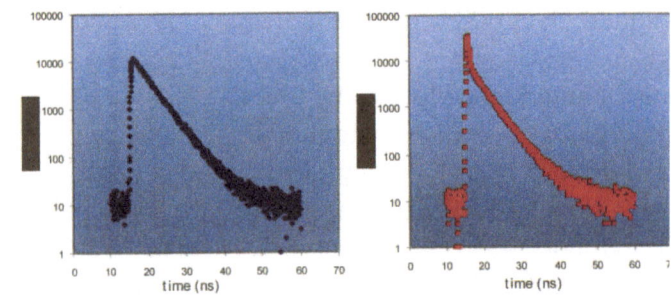

Fig.1. The normal decay of Fluorescein in PBS (10nM), 96 well black plate

Fig.2. The decay curve from a well containing Fluorescein and a highly fluorescent compound

- True bottom reading detection

Tecan have additionally added a further optical feature that extends the flexibility of the Ultra Evolution. The true bottom reading option retains the user-friendly function of "z" axis optimisation.

- Multi-Check QC Pac

With growing demand for traceability in detection devices, Tecan have developed the worlds first QC plate that supports the major detection methods available in Ultra Evolution.

Please visit the Tecan home page at www.tecan.com for more details or contact:
Joanne Woodward, Product Manager at Tecan Austria GmbH, Untersbergstrasse 1A, Grödig/SALZBURG, Austria, Tel: +43 6246 8933 x 137; Fax: +43 6246 8933 6137; Joanne.woodward@tecan.com

Date submitted: 23rd August 2002
Thermo Electron Spectroscopy

Thermo Spectronic
820 Linden Avenue, Rochester,
New York, 14625,
USA.
Tel: USA (585) 248 4000, Fax: USA (585) 248 4200
info@thermospectronic.com
www.thermospectronic.com

Specialty Keywords: **AB2, Aminco-Bowman, Spectrofluorometer, Fluorescence, Luminescence, Phosphorescence**

The Aminco Bowman Series II luminescence spectrometer (AB2) is a general purpose spectrofluorometer that can be applied to a wide variety of analytical problems.

Its main features include fast, sub-millisecond rate data acquisition and research grade optics that allow spectral bandpass settings of 0.5, 1, 2, 4, 8 and 16nm on each of its two monochromators. The AB2 has a signal to noise sensitivity of 900:1 (using the peak to peak noise calculation) or 2000:1 (using the root mean squared noise calculation).

A complete menu of built-in software applications and a long list of accessories make the AB2 a versatile instrument for research and routine applications.

A 150 watt continuous wave xenon lamp (for best sensitivity) and a xenon flash lamp (for phosphorescence measurements) can both be mounted inside the instrument.

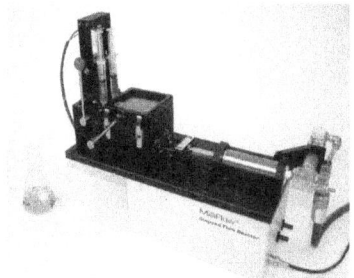

The AB2's major accessories include the MilliFlow™ stopped flow accessory for monitoring fast chemical reactions at millisecond rates (shown left, above) and two types of polarizers. The filter wheel polarizers (shown left, below) are for use in the visible spectrum and are less expensive then the AutoPolarizer (not shown) which has quartz prism polarizers for work in the uv-visible.

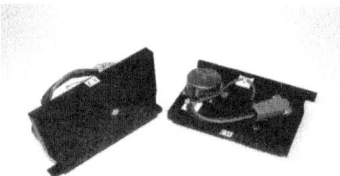

Software and sample handling accessories for the determination of intra-cellular ion concentrations are available. An optional T opticonfiguration can be used for fast dual emission wavelength work and for fast polarization measurements.

Instructions for Contributors

Scientists and workers in academia, industry or government employing fluorescence in their everyday working lives are invited to apply for entry in the *Who's Who in Fluorescence* annual volume.

The annual volume, edited by Chris D. Geddes and Joseph R. Lakowicz, publishes the names, addresses, contact details and a brief paragraph describing fluorescence workers specialities.

To apply for entry in the Who's Who in Fluorescence 2004 volume, complete the personal template (Word 2000 format) found at http://cfs.umbi.umd.edu/jf/ and e-mail to wwif@cfs.umbi.umd.edu no later than August 31st 2003. Unsuccessful entries, entries not conforming to the template format, or those received after the closing date will be returned without further consideration.

Contributors are asked to keep file sizes as small as possible by using appropriate standard picture formats, such as JPEG and TIFF etc. Alternatively, electronic versions can be submitted by post (CD) to:

Chris D. Geddes and Joseph R. Lakowicz
Editors: *The Who's Who in Fluorescence*,
The Center for Fluorescence Spectroscopy,
Medical Biotechnology Center,
725 West Lombard St,
Baltimore, Maryland, 21201, USA.

Galley proofs of entries will appear on the Who's Who website after the closing date. Contributors are asked to verify all details, with regard to *typesetting errors only*, within the 2-week period and return corrected proofs, preferentially via e-mail.

Personal half-page entries in the Who's Who in Fluorescence 2004 volume are free of charge. Further instructions and announcements will be posted on the website during the Who's Who entry collection period, January 1st – August 31st annually.

Fluorescence based companies may also submit a full-page company profile in the Who's Who in Fluorescence 2004 volume for a fee of $600.00 (black and white), $2000.00 (4-colour), prices subject to change. Full-page company templates may be found at: http://cfs.umbi.umd.edu/jf/ For colour images and high resolution images, companies are asked to contact the editors to discuss their requirements beforehand.

Institutions, academic research groups and centres of scientific excellence are also invited to submit full-page profiles for a fee of $250.00 (black and white), $2000.00 (4-colour), also using the company template found on the Who's Who website. Both company and institutional submissions are also to be submitted by August 31st 2003.

Further enquiries are to be directed to the editors at the above address or to: wwif@cfs.umbi.umd.edu